Intimate Universality

Local and Global Themes in the History of Weather and Climate

JAMES RODGER FLEMING,
VLADIMIR JANKOVIC,
AND
DEBORAH R. COEN

EDITORS

Science History Publications/USA
Sagamore Beach
2006

First published in the United States of America
by Science History Publications/USA
a division of Watson Publishing International
Post Office Box 1240, Sagamore Beach, MA 02562-1240, USA
www.shpusa.com

© 2006 Watson Publishing International

Science-History Studies on Atmospheres, Volume 1

Library of Congress Cataloging-in-Publication Data

Intimate universality : local and global themes in the history of weather and climate /
    edited by James Rodger Fleming, Vladimir Jankovic, and Deborah R. Coen.
        p. cm. — (Science-history studies on atmospheres; v. 1)
        Includes bibliographical references and index.
        ISBN 0-88135-367-1 (alk paper)
        1. Meteorology—History. 2. Climatology—History. 3. Climate and civilisation
    4. Science—History. I. Fleming, James Rodger. II. Jankovic, Vladimir, 1961–
    III. Coen, Deborah R.

√QC855.I58 2006
    551.509—dc22
                                                                    2006045098

Design and typesetting by Publishers' Design and Production Services, Inc.
Manufactured in the U.S.A.

Contents

Acknowledgments   v

Preface   vii

Introduction   ix

1

Intimate Climates, from Skins to Streets,
Soirées to Societies   1
VLADIMIR JANKOVIC

2

A Shift of View: Meteorology in John Herschel's
Terrestrial Physics   35
GREGORY A. GOOD

3

Mapping Meteorology   69
KATHARINE ANDERSON

4

Fog, Dust and Rising Air: Understanding Cloud Formation, Cloud Chambers, and the Role of Meteorology in Cambridge Physics in the Late 19th Century 93

RICHARD STALEY

5

Scaling Down: The "Austrian" Climate between Empire and Republic 115

DEBORAH R. COEN

6

Teaching the Weather Cadet Generation: Aviation, Pedagogy and Aspirations to a Universal Meteorology in America, 1920–1950 141

ROGER TURNER

7

The Struggle over Airways in the Americas, 1919–1945: Atmospheric Science, Aviation Technology, and Neocolonialism 175

GREGORY T. CUSHMAN

8

Global Climate Change and Human Agency: Inadvertent Influence and "Archimedean" Interventions 223

JAMES RODGER FLEMING

Notes on Contributors 249

Index of Names 253

Acknowledgments

Generous support for this volume was provided by the International Commission on History of Meteorology.

Thanks are due Neale Watson and his dedicated staff at Science History Publications/USA for seeing this volume through the production process.

Preface

These eight essays emerged from a meeting on the history of meteorology organized by the International Commission on History of Meteorology (ICHM) in the summer of 2004.[1] The venue was the baroque library hall of Kloster Polling in Upper Bavaria. Seventy-five participants from 23 nations made this, in the words of Gregory Cushman, "the largest gathering of historians of meteorology ever held." Although sessions were not without contention, an atmosphere of "scholarship and friendship" pervaded the meeting. Themes of "Intimate Universality" and the local and global dimensions of meteorology were freely discussed. Conversations continued over meals, during outings to the nearby historic mountain observatory at Hohenpeißenberg, after the evening concert, and at the farewell banquet held at the monastery.

Foundations for this volume were established in a session convened by Deborah Coen at the History of Science Society meeting in Cambridge, Massachusetts in 2003. Plans for the volume itself were hatched in the Polling Biergarten, after hours in the town square of nearby Weilheim, and during an impromptu hike on the Alpspitze.

It was in a snowfield on the mountain that Greg Good and Jim and Jamitto Fleming constructed a tiny "replacement" snowman (Figure 1) for Vlad Jankovic out of yellowish, grainy snow (Vlad had previously complained that a snowman he had made for his son following a rare snowfall in England had been stolen (!) from outside his door in Manchester). The yellow color of the snow comes from Saharan sand which is deposited during Genoan cyclones. So the little artifact made of natural material represents the hemispheric circulation, the Sahara expanses, and a gesture of friendship.

FIGURE 1   Miniature snow figure near the summit of the Alpspitze, July 2004 (map in background).

In offering this volume for your consideration, we extend our sincere wish for additional adventures in scholarship and friendship, perhaps in explorations in other fields of science and technology comprising and juxtaposing the intimate and the universal, the local and the global.

The Editors

NOTE

1. For details of the 2004 ICHM meeting, abstracts, and photographs, see http://www.meteohistory.org.

Introduction

The atmosphere is as intimate and personal as a breath of air in our lungs; it is also a global phenomenon with international political and economic dimensions. Writing histories of weather and climate in the twenty-first century means coming to terms with both the local and global dimensions of our modern lives. These histories must do more than trace the careers of meteorological instruments and concepts. After all, where is the weather? In Greek cosmology, the "aēr" in which human lives were immersed was distinct from the "aithēr," the realm of the gods. Today, however, atmospheric phenomena are at once unreachably distant and unavoidably near. On the global scale, the atmosphere moderates the planet's temperature through the natural greenhouse effect, allowing water to exist in vapor, liquid, and ice phases; it runs the fresh water or hydrologic cycle; it protects us from ultraviolet radiation and much space debris. Direct and scattered light produce halos, rainbows, and all the "glories" of the clear blue sky and glowing sunrises and sunsets. The atmosphere moves in a bewildering mixture of orderly and haphazard ways, from thousand-mile-long trade winds and westerly waves to swirling winter storms, hurricanes, tornadoes, gusts, and tiny unseen whirls. By comparison, the moon, the same distance from the sun, but without an atmosphere, is a harsh and lifeless object. On a human scale, our lungs encounter the global atmosphere through our alveoli, where the air combines with blood and enters into physiological rhythms. Since the air molecules circulate through the biosphere, scientists say that in each of our lungs right now (take a deep breath) there are

approximately one billion molecules that were breathed by any given historical figure in their lifetime: there are one to ten molecules of the Buddha's last
breath in each of your lungs right now. Unlike the airs of antiquity, atmospheric phenomena thus exist for us as abstract scientific objects at the same
time as they intrude all too tangibly on our health, mobility, and economic
well-being. The atmosphere is equally the property of experts and the public. It is a site of the encounter of the intimate and the universal.

What is "modern" about the modern sciences of weather and climate?
Modernity has been linked with globalization and integration. The history
of modern meteorology and climatology has been written as a change from
a local to a global view of weather.[1] While true to a first approximation,
such clichéd characterizations require a critical rethinking of timelines and
scales. Perhaps it is the age in which we live—marked by our daily juggling
of tasks that often have long-term and large-scale outcomes—that has made
us realize how ambiguous are the half-triumphalist and half-wishful buzz-
words like globalization, information, and networking. What does it mean
that a science is global or local? How local (or regional) was the natural
world in early modern times?[2] In what sense then, other than the rhetorical,
does modern meteorology claim to be global?[3]

INFRASTRUCTURES

Modern meteorology lays claim to universality first in its infrastructure. The
nineteenth century drive to standardize and coordinate world-wide weather
observations created a meteorological "synopticon." The science expanded
by exchanging instruments and enlisting observers, amassing data and comparing averages, charting airflows and sketching isolines, venturing forecasts and challenging hypotheses. There was a dose of Baconian naiveté in
much of this grandness and an element of disingenuousness in the international camaraderie epitomized by the creation of the International Meteorological Organization in 1873. Large-scale meteorology spread through
the arteries of empire: observations were carried by ship, railway, and telegraph, followed in the twentieth century by radio, TV, satellites, and the
Internet.[4]

The meteorology that evolved in this way is indisputably "big science."
As Steve Harris argues, sciences from meteorology to planetary astronomy
to natural history were shaped by the goals of the "long-distance corporations" that spearheaded them in the early-modern period. Whether their

goal was the promotion of trade, the propagation of the faith, or the investigation of nature, these nations and institutions enforced a site-oriented approach to soil, diseases, airs, and organisms. Local knowledge enabled surveillance, control, and domination, as is apparent in the premiums bestowed on environmental information by colonialists, clerks, conquerors and commercial traders.[5] Atmospheric science promised to pave Europe's path to the tropics by solving the problem of acclimatization and by casting a net over the atmosphere that ensured the movement of goods and cleared a path for domination. If weather cycles could be foretold, colonial rebellions triggered by agricultural crises might be suppressed by military force. The First World War made the stakes clear: the trajectory of artillery shells, the course of airplanes, and the diffusion of poison gas all required new forms of meteorological intelligence. Today's militaries likewise aspire to fight and win wars on a global battlefield that encompasses the atmosphere, oceans, polar caps, and near-space environments. The strategic and tactical needs of an all-weather air force, a nuclear navy, and the targeting requirements of offensive and defensive missile systems all demand detailed geophysical knowledge.[6]

The market contributed to meteorology the imperative to leave nothing to chance. A dry season or a single storm could throw a market into a tailspin. Weather prediction was a calculation of financial risk on which fortunes could be made or lost. Managing the environment involved attempts by scientists to understand and predict weather patterns, delineate climatic zones and vicissitudes, and probe the nature of storms and the composition of air. Government bureaucrats, physicians, merchants, farmers, and insurers then attempted to appropriate and apply knowledge from the global synopticon for planning purposes, both domestically and in the colonies, including trade routes, health policies, agricultural practices, and military campaigns.

In this quest to manage the weather, global imagery emerged early: "The Meteorological Society . . . has been formed, not for a city, nor for a kingdom, but for the world," wrote John Ruskin in 1839. "It wishes to be the central point, the moving power, of a vast machine, and it feels that unless it can be this, it must be powerless; if it cannot do all, it can do nothing."[7] Indeed, it could not do all: just three years later the society dissolved. Whether through shifts of technique and instrumentation or through the abandonment of observing stations and records, the grandiose projects of modern meteorology have often been cut short. The infrastructure of globalization itself thus evinces a counter-tendency towards fragmentation.

ONTOLOGIES

For modern scientists, reasoning about the atmosphere has often meant conjuring with vast elemental forces, seeking explanations at superhuman scales. If the Moon's gravitational force could raise immense ocean tides, it was logical to assume that it could also be the cause of periodic weather patterns and large-scale disturbances such as atmospheric waves. The rotational force of the Earth itself and the sun's heat captured over the entire globe were imagined to be condensed in the whirling, swirling, and damaging winds of gales. The electrical nature of lightning inspired naturalists to wonder if the electrical fluid triggered earthquakes, spread diseases, lit up Northern Lights, and might be used to annihilate humidity in the atmosphere.

It took ingenuity to tie these cosmic theories to the micro-scale of the emerging sciences of chemistry, thermodynamics, and electromagnetism. When meteorologists imagined the atmosphere as "a vast laboratory," they effectively erased distinctions among the electrical, chemical, and physical, between "outdoor" and "laboratory" natures. This fusing of conceptual space opened the door for studies of clouds and cloud chambers, maize and malaria, hot air masses and hot air balloons, agriculture and aviation to be seen as part of a common enterprise. The merging of big science and small drew attention agents on the molecular scale. John Tyndall (1820–1893) experimented on carbon dioxide in the free air and in human lungs. The geographer Ellsworth Huntington (1876–1947) thought it was ozone that accounted for peak mental and physical activity. In 1946, Bernard Vonnegut (1914–1997) identified silver iodide as a trigger mechanism that might allow widespread modification of the weather. Sulfuric and nitric oxides were central in acid rain controversies; and protecting ozone in the stratosphere became the goal of the Montreal Protocol and its successors. Today, everyone's environmental molecule of choice, carbon dioxide, has become an international symbol of human intervention in the climate system, signifying both affluence and apprehension.

Most scientists today believe that the microphysical processes active in a cloud ultimately result in macroscale weather phenomena. Yet researchers do not have the ability to characterize the background concentration, sizes, chemical composition, and time variation of the basic cloud constituents: aerosols, ice nuclei, and liquid water droplets. Other factors such as chemical reactions, turbulence, and radiation complicate the picture even more. Numerical modelers use parameterizations with little theoretical basis that cannot properly account for the complex microphysical processes occurring at sub-grid scales with great spatial and temporal variability.

The history of meteorological ontologies, like that of the science's infrastructures also reveals a tendency towards fragmentation. The new quantitative, physical models of meteorological processes of the nineteenth century highlighted their irreducibly local character. Resting their claims on new physical theories, researchers concluded that storms in the northeastern United States were in fact different in character from those in Europe. Analogously, others argued that the Föhn winds of the Alps were of local rather than tropical origin, their character shaped by regional topography. Centers of excellence emerged at meteorological institutes that specialized in particularly noteworthy local phenomena. Interested in the health and well being of their citizens, governments launched studies in the microclimates of workspaces and urban environments in which several levels of scales came under scrutiny in the same way as did the outdoor spaces. The late nineteenth century reforms of the domestic realm, including ventilation, heating, and dress reform, extended meteorology's terminology to encompass the intimate space between skin and clothing, between clothing and the walls of the house. Appropriately down-scaled, the meteorological idiom became a tool in charting the moral topographies of modern living and their role in the political economy of nations. So it was that Britton Armstrong Hill (1818–1888), a prominent St. Louis attorney, urged that meteorology be pursued not for its own sake but as a science of "ventilation and hygiene, so that the whole of this globe of ours may, in course of time, be made inhabitable and healthy."[8]

On these smaller scales, atmospheric phenomena have become clues to anthropological investigations of identity. The weather lore of Norwegian fishermen, the rain rites of Tanzania's Ihanzu people, and the emotional attachments and municipal politics of American snow—all these discourses of weather can be seen as constructed through day-to-day, season-to-season activities.[9] Even in the most recent theoretical and operational meteorology, locality enters not only for the simulation of small-scale phenomena, but also in the value accorded to personal experience of atmospheric changes, whether of the tropics, mountains, deserts, or lakes.[10] Similarly, "local climates" draw the interest of those who seek to naturalize the borders of new nation-states or bear out the autonomy of regions, cities, even neighborhoods. Based on investigations of the local 'air,' regions have been vaunted for their 'bracing' or 'ennervating' qualities or condemned for their 'marshy' or 'suffocating' powers. Thus in the colonial age, the tropical climate became inextricably associated with the character of the tropics' inhabitants.

In the end, however, the local and global versions of meteorological history are not incompatible, and this volume seeks to bring them into productive exchange. Neither trajectory, from "local to global" or from "global to local," can be considered inevitable. Instead, we ask what drove scientists, officials, and publics to draw the horizons of their view as they did.

NARRATIVES

Arranged chronologically, the essays in this volume span three hundred years of meteorological history. It was in the eighteenth century that humans first became cognizant of their influences on nature on large spatial and temporal scales. Developments in scientific forestry and hydrology in this period, for instance, signaled the desire to bring unruly natural environments under human control. It is into this transformative period that Vladimir Jankovic leads us in the first essay in this volume. The Enlightenment physicians who are his focus retained elements of an older environmental determinism, seeking to understand the effect of particular climates on the health of their inhabitants. In this vein, medical wisdom of the day warned that bad air was to blame for much of human disease. Yet these physicians also expanded their view of climate in a characteristically modern way. Paralleling the population shift to urban centers, the rise of reading clubs and other institutions of the public sphere, and the emerging bourgeois focus on domestic life, eighteenth-century physicians developed a new concern with the airs of urban interiors. Expressing a typically modern fear of the debilitating effects of civilization, physicians sought to mitigate the dangers of indoor climates. Jankovic shows that the Enlightenment bid to engineer the environment unfolded on the small scale as well as the large, bringing new forms of discipline to the intimate conduct of human bodies.

The large-scale conquest of nature, meanwhile, depended on new modes of collecting natural knowledge on a global scale. Imperialist expansion put in place new networks of scientific observers, the coordination of which is addressed from different angles by Gregory Good and Katharine Anderson. Good uncovers the little-known meteorological researches of John Herschel (1792–1871), famous for his astronomy and philosophy of science. Herschel was remarkable in the way he unified the diverse and potentially conflicting ambitions of the young science of meteorology. Meteorology, as he saw it, was an empirical science that required precise measurement and intimate, first-hand knowledge of local airs. Yet Herschel was equally convinced that meteorological phenomena were subject to universal laws, accessible through induction and the testing of hypotheses. In Good's portrait, Herschel stands as a symbol of meteorology's multiple potentials in the nineteenth century: a discipline equally committed to a Romantic experience of nature and to a bureaucratic model of managing big science.

Beginning in the mid nineteenth century, the spread of the telegraph enabled a new level of global scientific communication about the weather. The new scale of weather observation raised more questions than it answered. How did atmospheric phenomena travel? Did the same laws govern weather in the arctic and the tropics? Even picturing the weather on a global scale proved challenging. Focusing on several competing solutions to the problem of synoptic (large-scale) weather mapping, Katharine Anderson shows that these various schemes presented the relationship between local and global knowledge similarly. On one hand, representations of small-scale details on these early maps highlighted the importance of disciplined, dutiful, and accurate local observers. On the other, the images in their entirety implied that local knowledge became valuable only in a global context of international coordination. Interpreting these representational conventions, Anderson depicts nineteenth-century weather prediction as an activity that highlighted the growing gaps between popular and expert knowledge, between personal experience and collective calculations.

Laboratory experimentation was yet another means of elucidating the relationship between local and global knowledge of weather in the nineteenth century. When laboratories dedicated to precision measurement first came to dominate the field of physics in the mid nineteenth century, there was no divide at all between the disciplines of physics and meteorology. Field and laboratory investigations of atmospheric phenomena continued to be conducted alongside each other. As Richard Staley demonstrates, scientific instruments played a key role in mediating the shifting relationship between the field and the lab. Instruments helped bring outdoor phenomena indoors, contributing to what Peter Galison and Alexi Assmus have called the tradition of "mimetic" experimentation.[11] At the same time, instruments made it possible to measure parameters relevant to laboratory investigations directly in the field. Staley focuses on the case of the cloud chamber, an instrument that was used both to mimic cloud formation and to analyze the properties of fundamental particles. As an object embedded in the practices of both laboratory and field science, the cloud chamber was caught up in debates over

FIGURE 1   Carl-Gustaf Rossby with a rotating tank used for studies of atmospheric motion, 1926–27. Felix Exner conducted similar experiments in 1923. Image ID: wea01501, Historic National Weather Service Collection.

the divide between "natural" and "artificial" phenomena. As Staley shows, these deceptively simple boxes helped relate the intimate particulars of the weather at a given place and time to the universal abstractions of fundamental physical theory.

As meteorology struggled to form its own disciplinary identity at the turn of the twentieth century, weather scientists most often modeled their field on physics, as a science of universal, deterministic laws. Robert Marc Friedman has traced this aspiration in the work of Vilhelm Bjerknes and the highly influential school of meteorology he founded in Bergen, Norway. Driving the universalizing vision for meteorology were broader developments in the early twentieth century, including the battlegrounds of the First World War and the urgent need for weather forecasts for pilots, followed by a renewed enthusiasm for international scientific collaboration in the interwar period. But not every successful center of meteorological research followed this universalizing pattern. In her essay, Deborah Coen describes the "down-scaling" of weather science in Vienna in the 1920s, after the city's de-motion from imperial metropolis to the battered capital of a defeated rump state. In the process of developing a climate science relevant to this small nation, Austria's meteorologists found themselves explaining for the first time what made a local climate "local." Attention to the relationship between atmospheric phenomena at different scales was not only a source of scientific innovations. The new prominence of weather science in Austria at this time also furnished a vocabulary which helped non-scientists—from novelists to political theorists—describe competing visions of the nation and of the amorphous entity increasingly identified as "Central Europe."

While Coen recounts meteorology's fate in one of the Great War's losers, Roger Turner explores its history in the victorious United States. Turner traces the transformation of American meteorology led by Bjerknes' Swedish disciple Carl-Gustaf Rossby. Bergen-school methods were not simply transplanted to the New World. Instead, developments were driven by the growing needs of aviation, an expanded upper-air observation network, and the outbreak of World War II, the last resulting in the training of thousands of weather forecasters after 1940. From this pedagogical regime, which taught forecasting as an application of universal physical principles and instilled a military ethos in its students, emerged the contours of postwar American meteorology. The goal of making meteorology a respectable, useful, and universal science in the U.S., Turner shows, had a highly localized, context-dependent impetus.

This universalist aspiration had practical as well as theoretical consequences. Gregory Cushman argues that meteorology was as much an "imperial" science in the twentieth century as it had been in the nineteenth. With this designation Cushman points both to meteorology's dependence on neocolonial power structures and to the tendency of its major schools to establish "colonial" outposts on foreign ground. His essay in this volume centers on the development of meteorology in the service of a burgeoning aviation industry in Latin America and Canada up to the end of World War II. Intriguingly, the global aspirations of Euro-American meteorology were defeated in the tropics, where Bergen-style air-mass analysis was replaced by techniques sensitive to the local effects of sun and mountains. The United States' effort to forge a "Pan-American" regional identity through a shared network of weather science was frustrated by political divisions and by the heterogeneity of the natural phenomena themselves. In a Latin America riven by the legacy of colonialism, the choice of the scale at which to study weather was highly politicized, just as it was in post-Hapsburg Central Europe.[12]

Today, the question is not just at which scale to study the weather but at which scale to begin to change it. As Jim Fleming cautions in the final essay of the volume, history should make us wary of technocratic projects to control the climate. Fleming traces the convoluted development of our current understanding of anthropogenic climate change, bringing the history of human influence on the climate—both inadvertent and intentional—to bear on questions of ethics and public policy. The twists and turns this science has taken even in the recent past suggest that we would be foolish to claim omniscience today. If indeed a "tipping point" in the global climate system is imminent, can changes in collective lifestyle avert an unprecedented and potentially catastrophic disaster? Will mitigation and adaptation be enough, or will technical elites attempt to "fix" the climate through geoengineering? Isn't it better to make science and public policy with a knowledge of history?

> JACK, "Charming day it has been, Miss Fairfax."
> GWENDOLEN, "Pray don't talk to me about the weather, Mr. Worthing. Whenever people talk to me about the weather, I always feel quite certain that they mean something else. And that makes me so nervous."
> JACK, "I do mean something else."
> GWENDOLEN, "I thought so. In fact, I am never wrong."[13]

This book is about meteorology as a tool and an artifact of modernity. Like all talk about the weather, meteorology is often done with something else in mind. It is a science of no place and every place, done both routinely and with an increasing sense of urgency. Through its history we witness confrontations with the randomness of material life across cultures, spaces, and epochs. More immediately perhaps than other experts, atmospheric scientists are charged with the rational management of natural resistances that often thwart the proceedings of everyday life. From the most intimate personal details to the most volatile matters of geopolitics, the atmosphere underpins and undermines our efforts at planning, rationalization, and control. Meteorology is a way of hedging our bets against the greatest natural given outside our doors and the most violent variable in human affairs, a given that is as banal as it is brutal—the grand nuisance of being thrown into the weather. In this struggle, we are all in it together.

NOTES

1. This is true primarily in the case of older, popular literature such as Bernard Ashley, *Weather Men* (London: Allman, 1970); H. Howard Frisinger, *The History of Meteorology to 1800* (New York, 1977; Boston: American Meteorological Society, 1983); and Hans Gunther Korber, *Vom Wetteraberglauben zur Wetterforschung* (Leipzig, 1989), yet each of these volumes has its value and appeal.
2. *Reading the Skies: A cultural history of English weather, 1650–1820* (Chicago: University of Chicago Press, 2000); Dane Kennedy, "The Perils of the Midday Sun: Climatic Anxieties in the Colonial Tropics," *Imperialism and the Natural World*, John M. MacKenzie, ed. (Manchester and New York: Manchester University Press, 1990), 118–40; David Livingstone, *The Geographical Tradition: Episodes in the History of a Contested Enterprise* (Oxford: Blackwell, 1993); Nicolas A. Rupke, ed., *Medical Geography in Historical Perspective*, Medical History Supplement No. 20 (London: Wellcome Trust, 2000).
3. Robert S. Spich, "Globalization Folklore: Problems of Myth and Ideology in the Discourse of Globalization," *Journal of Organizational Change Management* 8.4 (1995): 6–29; Manuel Castells, "Toward a Sociology of the Network Society," *Contemporary Sociology* 29 (2000): 693–99. Elisabeth Crawford, Terry Shinn, and Sverker Sörlin, eds., *Denationalizing Science: The Contexts of International Scientific Practice* (Dordrecht: Kluwer, 1993).
4. On meteorology in the nineteenth century see, for example, Gisela Kutzbach, *The Thermal Theory of Cyclones: A History of Meteorological Thought in the Nineteenth Century* (Boston: American Meteorological Society, 1979); James Rodger Fleming, *Meteorology in America, 1800–1870* (Baltimore: Johns Hopkins University Press, 1990); and Katharine Anderson, *Predicting the Weather: Victorians and the Science of Meteorology* (Chicago: University of Chicago Press, 2005).
5. Steve J. Harris, "Long Distance Corporations, Big Sciences and the Geography of Knowledge," *Configurations* 6 (1998): 269–304; Richard H. Grove, *Green Imperialism: Colonial expansion, tropical island Edens and the origins of environmentalism, 1600–1860* (Cambridge: Cambridge University Press, 1995).
6. James R. Fleming (ed.), *Historical Studies in the Physical and Biological Sciences* 30, no. 2 (2000), special theme issue on geophysics and the military.
7. John Ruskin, "Remarks on the Present State of Meteorological Science," *Trans. Meteorol. Soc. Lond.* 1 (1839): 56–59.
8. Britton Armstrong Hill, *Liberty and Law, or, Outlines of a New System for the Organization and Administration of Federative Government* (St. Louis: G.I. Jones, 1880).
9. Robert M. Friedman, *Appropriating the Weather: Vilhelm Bjerknes and the Construction of a Modern Meteorology* (Cornell University Press, 1989); D.T. Sanders, "Rainmaking, gender and power in Ihanzu, Tanzania, 1885–1995,"

Ph.D. dissertation, University of London, 1997; Bernard Mergen, *Snow in America* (Washington, DC: Smithsonian Institution Press, 1997).

10. Sarah Strauss and Benjamin S. Orlove (eds.), *Weather Climate, Culture* (Oxford: Berg Publishers [Oxford International Publishers, Ltd.], 2003).

11. Peter Galison and Alexi Assmus, "Artificial Clouds, Real Particles," in *The Uses of Experiment*, ed. David Gooding, Trevor Pinch, and Simon Schaffer (Cambridge: Cambridge Univ. Press, 1989), 225–74.

12. On the relevance of "scale politics" to the history of science and medicine, see Gregg Mitman, Michelle Murphy, and Christopher Sellers, "Introduction: A Cloud Over History," in *Landscapes of Exposure*, *Osiris* 19 (2004): 1–17, and Nicholas B. King, "The Scale Politics of Emerging Diseases," *idem*, 62–76.

13. Oscar Wilde, *The Importance of Being Earnest*, Act I, Project Gutenberg EBook #844 (1997), http://www.gutenberg.org/dirs/etext97/tiobe10h.htm

CHAPTER 1

Intimate Climates

From Skins to Streets, Soirées to Societies

VLADIMIR JANKOVIC

'What does a fish know about the water in
which he swims all his life?'

—Albert Einstein[1]

When in February 1737 philosophy professor William Cockburn of the Edinburgh University died from a sudden "effusion of blood from the lungs," it was reported that the sun assumed an eerie scarlet tinge caused by a haze that made the air oppressive and exhilarating by turns. The atmosphere was halted in a ghastly calm as if before a cataclysm; the city barometers fell to the point where they would, if such were possible, indicate deluvial rains. The same afternoon, Cockburn's colleague, the distinguished Dr. Archibald Pitcairne, experienced a paralyzing attack of weakness following a profuse nose bleeding. In the city infirmary, physicians could not come to terms with a series of hemorrhages that crippled a dozen patients for no obvious reason. Commenting on the episodes, George Sigmond, an early Victorian expert on *materia medica*, thought that the simultaneity of afflictions pointed to a common meteorological cause, suggesting that the instances in which the weather affected human health in the above way, if recorded, would throw "some light on subjects which ought not to escape the most anxious investigation."[2] Sigmond had in mind medical meteorology, a genre devoted to understanding the relationship between the

1

body and its socio-physical surroundings that in the last decades of the eighteenth century consumed the energy of the ever-growing numbers of practitioners seeking to resolve an impasse created by the ineffectual heroic treatment and the perplexing pharmacopoeia of the medicinal market. Groping for ways to tackle epidemics and chronic disease, such practitioners spread the word of medical meteorology across Europe, the Americas, India, and the Mediterranean. The period experienced an unprecedented zeal to explore an ever-widening relationship of airs to health and an interest in seeing human (and national) welfare as a matter of geographical situation, climatic pattern, and environmentally defined civil engineering, domestic architecture, comfort technologies, clothing, and sport.[3]

In recognition of these broad developments that marked theory and social practice, it is argued that, some time before the year 1800, the European and American medical profession created a field of "environmental medicine" (Ludmilla Jordanova), "medicine of climates and places" (Michel Foucault), the "environmental paradigm" (David Arnold), or "environmentalism" (James Riley).[4] Such terms are meant to describe a wide-ranging interest in the quantitative investigation of the properties of airs, gases, atmospheres and climates, both in and outside the laboratory and the lungs. The period is associated with researches in eudiometry, medical topography, altitude physiology, medical pneumatics, gas chemistry, and climatotherapy. A large number of people, both educated and lay, showed an unusual awareness of the so-called atmospheric constitutions.[5] Governments expressed urgency in finding the solution to the effects of foul air generated onboard ships and in prisons, hospitals, and other public buildings. Practically everyone knew that malaria, miasmas, and the potpourri of miscellaneous mephitic vapors were the prime suspects in the epidemics of cholera, fevers, and typhoid. There was no question, among the educated public, about the health risks associated with rebreathing, overheating, sudden thermal changes, and accidental inhalation of vapor from marshes, macerating pits, coffins and mines.[6] Urban planners and health officials showed enlightened concern about the layouts of suburbia and the location of cemeteries to avoid public exposure to the proven epicenters of disease.[7] Dissenting philanthropists worried about "the relationship between health and a person's environment and personal status," while explorers, colonial doctors, and navy officers reported on the effects of pestiferous tropics and bone-numbing cold.[8] Back home, changes in domestic architecture accommodated heating and ventilation requirements. The climatological citizenship that such developments inaugurated entailed a putative polity—a climatological state—whose

2

boundaries reflected the social imagery of atmospheric powers held to structure human health and history.[9]

In this essay, I'd like to draw attention to a dichotomy that enabled this form of virtual citizenship: one between the indoors and the outdoors. I wish to draw attention to this dichotomy because its (important) role in the systematic analysis of the weather as a "milieu" has rarely been a subject of meteorology and, by extension, of the history of meteorology. Perhaps rightly so: such weather is a condition, not a measurable entity. It is at once private and public and it can become a subject of science only when certain circumstances are met. Before it becomes so, however, the weather is encountered in a variety of ways. In trade, war, and politics it may be regarded as a frustrating resistance to human ambition; in agriculture, insurance and medicine it is a risk factor; in art a symbol of transcendental potency, in religion of God; in daily lives it is rarely more than a nuisance. As scholars, we generally know little about these weathers and yet it is an incomplete history of meteorology that neglects their existence simply because they seem difficult to scrutinize. Instead, I would suggest that these are precisely the conditions that give meaning to meteorology as a science. One learns important things about the historical meandering of meteorological interests if one sees them as related to the social and somatic attitudes toward the weather as a milieu (or toward the absence of it as such). Seeing the weather *before* it is assembled and appropriated as a subject of a shared discourse based on a rational discussion and exchange of comparable data—as in some way a boundary object straddling conceptualizations—is like seeing a soil before it is cultivated. What is cultivated depends on the nature of soil. In this essay, accordingly, I'd like to draw attention to the weather as this soil, before it is cultivated as a subject of medical meteorology, one of its many systematic appropriations.

In doing so, the main emphasis of this article is on the social, moral and medical dialectics between the indoor and outdoor spaces as they were described in the literary and medical culture of eighteenth-century Britain. I suggest that the crossing of the boundary between the two was a sociomedical affair constructed in terms of the "exposures" to atmospheric and man-made conditions affecting the health, comfort, and status of people in Britain and elsewhere. Particularly important in this analysis is the role of the indoors, which so far has received a marginal treatment within environmental history, still less in the history of medicine and meteorology. We possess no studies on bathroom monsoons, attic drizzles, or the oven Scirocco. The existing histories of meteorology make it natural to treat the weather as a subject which was to be found only outside the enclosed space of controlled comfort—i.e. outside the interiors in which people live, work, sleep, and entertain themselves for no less than 80 percent of their lives. It has also become natural to forget the trivial fact that the weather would hardly ever merit a notice were it not seen from "somewhere else." It would not make sense without an implied comparison to an enclosed atmosphere of comfort. Outside architectural microclimatology, toxic exposures, and occupational health, climatological issues are very rarely discussed in relation to small, limited volumes of air and yet—one may argue—these were precisely the volumes about which people cared and considered the real subject of their (intimate) meteorologies. When the ambers died, it was the chill of the bedroom, not the thermometer, that measured the frost outside.[10]

But how to approach this unwritten history? For one, the shelter and the weather had constructed each other since the beginning of purpose-built habitation. By the nineteenth century, this dialectic produced a weather discourse that made no quarrel about treating the indoor atmospheres with equal, sometimes greater, concern than the outside rain and wind. Most obviously, the interest in "air" was often made possible by an interest in "bad air." The indoors bred damp, dusk, and dust of a different character than those found outside. One defined the shelter by an exclusion of what one perceived intolerable. And one defined what was intolerable by what one believed could be excluded. In an important way, the rise of "indoor climates" followed on the heels of the rise of the European and American culture of comfort and, as a consequence, that of dis-comfort. The discomforts of kitchen fires, culinary emanations, fireplace drafts, and soggy corridors that sported moss on their steps became paragon evils supporting the case for domestic hygiene and early public health engineering, architectural alterations, and medical police measures. It was merely a matter of time before the medical reformers began to justify the urban clean-up in small-scale terms. They, for instance, spoke of ventilating, deodorizing, and washing the towns as if they could be treated like privies, houses, barracks, hospitals, or even bodies. This means, among other things, that outdoor rights in modern climatological states followed the legislative steps previously made to ensure the rights of comfortable interiors.[11] But what did such trends have to do with meteorology?

A case could be made that a great portion of quantitative meteorology before the 1830s derived from medical concerns about the pathology of early modern airs. "Meteorological" observations prior to the 1830s—i.e. before European, American, and colonial institutions hijacked the science as a

paragon of scientific method and public enterprise and before it became im-
bricated in the forecasting controversy—owed their thoroughness to medical
concerns over the nature of indoor and outdoor stress, whether in the form
of atmospheric malaria, miasmas, drafts, smoke, winds, heats, droughts, or
electrical states of the air. Much energy went into thinking about how to
replicate the qualities of atmospheric air in domestic and urban spaces, probably as much as went into traveling to places such as the Mediterranean in
which such air naturally replicated the equability and thermal properties of
the cozy home back North. Caring about the body had often to do with caring about the airs that enfolded it.

This relationship could be illustrated by the fact that the eighteenth-
century medics represented the single largest proportion of professional people engaged in taking instrumental readings of the weather.[12] In Britain, early
networks began with eminent physicians like James Jurin and Thomas Short.
During the eighteenth century, continental authorities instituted medico-
meteorological observatories to learn about conditions during epidemics.
Colonial expansion and trade travel inured European doctors to the unusual heats that alerted them to the necessity of correlating the atmospheric
parameters with the progress of acclimatization. Early in the nineteenth century, three out of five founders of the Meteorological Society of London
(1823) were doctors: George Birkbeck, Henry Clutterbuck, and Thomas
Ignatius Maria Forster (the two others were the noted meteorologists Luke
Howard and John Frederick Daniell).[13] James Tatem, who lobbied for its
foundation, explained that the society would be in charge of a nationwide
archive of weather journals. "From registers thus formed, the meteorologist
would received information and pleasure, the man of science amusement,
and the valetudinarian benefit, by being able to select a residence where the
climate suited his constitution—a thing of no small consequence, and no little difficulty in the variable temperature of our native isles."[14] Tatem's contemporaries took it for granted that meteorology made no sense outside its
application: "nothing could be more jejune and uninteresting," wrote medical topographer John Hennen, "than a protracted enumeration of the daily
variations of the atmospheric temperature, weight, and moisture, or the different shiftings of the winds, if the person who describes such occurrences
does not deduce from them some practical information, by marking the effects which they produce upon the health of man, and upon the face of nature."[15] The interest in weather, as is perhaps only too obvious, had an
important personal dimension. To what extent is this also true of medical
meteorology as a science?

To try to answer this problem, I will first sketch the state of medical meteorology around the year 1800 and briefly look at its scholarly and experimental origins. I will then single out the importance of the physiological investigation concerned with the so-called nosopoetics of air, i.e. the air's physical properties thought to have a pathological influence on the healthy operation of all bodily functions. Following this will be an excursion into the broader cultures of Enlightened "airmindedness" and its role in raising an alertness to the externally caused diseases associated with luxury and "artificial" modes of life. The latter, it is argued, cut through the notions of urbanization, nervous sensibility, and Enlightened sociability to chart the narratives that I think should claim our attention as the sources of an indoor-inspired pathological meteorology. A conclusion offers thoughts on the relationship between intimate climates and the invention of the environment. I should however note that no space could be allotted to a more focused presentation of air chemistry, pneumatics, and medical topography, all of which are relevant to the topic but have been treated authoritatively by others. Finally, I will also leave out the developments in the construction of prisons, ships, hospitals, and workshops which, again, had very much to do with the understanding of indoor climates. My primary concern is with the medical cultures of domestic atmospheres.

ENLIGHTENED AIRMINDEDNESS

The interest in climate as a spring of social life had a strong pedigree. The founding fathers of modern climatological states were the anthropologists of atmospheric *longues-durées*. Eighteenth-century academics, philosophers, wits, and political commentators interested in the comparative study of European and world cultures produced a considerable opus on the ways in which the average weather determined the character of races, endemicities, and forms of government. The Mediterranean air created Italian *belcanto*; the polar wind was responsible for the monosyllabic, baritonic barbarism of the Northerners. The fever of the Norfolk fen could not be found in high Derbyshire; cretinism of the Alps had its origins in the smells of local manure and the Atlantic storms imbued the mid-latitude peoples with independence, decency, and democracy. Enthusiastic as were Montesquieu's and Falconer's or scathing as were Diderot's and Boulanger's, these musings broached the possibility that social affairs hinged on an agency external to conscious forms of political action, moral reasoning, legislative process, or military strategy. Instead, political theorists thought that life mimicked extrinsic (now made intrinsic) forces of soil, water, topography, and climate. Racial, medical, and political lives were seen as steered from outside of their cultural reproduction, except that this outside was now recognized as the *sine qua non* of the subjects' multiple destinies.[16]

Can I suggest that the analyses of health as a result of these direct influences of airs on individuals and populations—atmospheric or "domestic"—emerged as the principal new development within eighteenth- and nineteenth-century Hippocratic medicine? Leading lights in the tradition showed that the so-called "nosopoetic" qualities of air—distinct from its miasmatic or malarial ingredients—could trigger a disease regardless of the presence of putrefying substances, because "the changes that happen in the Atmosphere which surrounds us, and the Air we breath, constitute the general, external causes that affect Health."[17] Dampness was *directly* responsible for colic, fluxes, and consumption; the inflammations of throat and eyes, rheumatic pains in the teeth and face, and coughs were, again, *directly* caused by cold air. In these ratiocinations, the non-miasmatic forces in the air's mechanical, hygrometric, and thermal properties acquired an unprecedented power over life. And while there is no doubt that earlier generations experienced the same forces on a daily basis, the novelty of the Enlightenment approach was in its comprehensive sweep that affected not only medical thinking, but also contemporary ethics, political theory, and the many bio-cultural embodiments of the Enlightenment bourgeoisie. On the mundane level, the practices related to mitigating these influences had a more lasting influence on the rise of an "environmental" outlook than chemical and epidemiological investigations.[18]

The period documents give an impression of genuine surprise regarding the stunning potential of the new nosopoetic properties of the air. Naturalists in particular dwelled on the oxymoronic notion of the air's weight, relentlessly pondering over trivia such as that the atmosphere, at any given time, squeezed the body with over 32,000 pounds of weight, even as no one noticed it. The air was heavy, "heavier than you imagine,"[19] and its pressure changes mimicked those of the air pump in which the animal swelled, constricted, absorbed, and released gases. The air pump illustrated the need for room-drafts to reduce the chimney smoke. Ben Franklin explained that if a fire required a supply of air, and if a fire continued to burn in a room with tight-fitting wainscoting, doors, and sashes, then the room "must soon be exhausted like the Receiver of an Air pump, and no Animal could live in it."[20] Claustrophobia could only be aggravated by the air's unsavory content that was invisible to the eye but more than offensive to the nose and the lungs. The list of these ingredients made by Rev. Stephen Hales, the air-discover and ventilation expert, included water, dust, salt, oil, buttery particles, volatile smells, vegetable vapors, flower perfumes, sweat droplets, volcanic ash, insect eggs, maggots, coal and nitrous smokes, odors of fermenting liquors, mine fumes: all these over and above the terrifying and inscrutable miasmas. On his calculation, if a body released an inch of sweat in a month, then an acre of human skin—made up of 2904 individual skins—would churn out a good 71-foot-high layer of pestiferous stuff—a fact worthy to be considered by hospital architects, ship-builders, and city planners.[21]

The important challenge that from early on faced the nosopoets of air was to find the mechanism whereby the air regulated animal economy. The standard discussions repeated what was known about respiration, intestinal absorption and release of gases, perspiration, and cutaneous transmission. The air was understood as a key principle of metabolism controlling a healthy operation of organs simply by "being introduced into all the bodies that have any kind of life."[22] But how did the body "know" how to respond to the external stimuli? Healthy respiration posed no problem in this respect, but different neuralgic reactions to atmospheric changes presented a difficulty because they implied a differential susceptibility. Medical vitalists identified this as the faculty of sensibility.[23] Especially relevant in developing this idea were the medical professors in the universities of Montpellier and Edinburgh. The French physicans spoke of "the general external organ"—involving the skin, the integuments and the tissue under the skin—which transmitted outside influences to the brain and the epigastric system: "it was like a gate wide open to all external impressions." This made sense of the Galenic *ingesta* and helped overhaul the doctrine of non-naturals from a new point of view.[24] But vitalists' notion of sensibility also enforced a broad "environmentalist" slant of French hygiene and medical police, as it tied the inner organization of the body with the social-cum-climatic circumstances in which it found itself. The Montpellier position thus helped early hygienists "to claim as part of their purview *any substance*, general environmental condition, or activity that contributed to the maintenance of the life of the organism (all *ingesta*), formed its milieu (air, water, climate, the Hippocratic "places"), or conditioned its own response (exercise, work, habits). Within such a rubric, *virtually anything* could be said to impinge on health or sickness and therefore to fall under the authority of the doctor."[25] It doesn't come as a surprise, then, to hear two recent medical historians claim that in France during the 1750s, an "obsessive concern [with] air quality" thrived on the presumption that the "air was the source of virtually *all* diseases," a testimony to scepticism about an "internal cause of disease," that, more generally, fostered an interest in non-naturals and "the external causes of disease."[26]

Sensibility and sympathy were the keystones of Scottish physiology and medicine. So was research into the physiology of air.[27] The sensibility doctrine enjoyed a consensus among the star-studded faculty of the University of Edinburgh (Robert Whytt, William Cullen, Monro Secundus, John Hunter, John and James Gregory) who pursued a Scottish crypto-vitalism that underlined the importance of Galen's non-naturals like air and diet in therapy and regimen.[28] Here sensibility was a stimulant of all somatic activity operated by means of the nerves and controlled by diet, exercise, climate, sleep and passions.[29] Sensibility became a catchphrase of the whole of Scottish Enlightenment. Importantly, it was defined as a sensibility *to* (stimuli) over and above the sensibility *of* (the nerves and the organism). For example, when David Hume, Lord Kames, and Adam Smith imaginatively examined the life of savages, they reckoned that their "insensibilities" *to* the plight of others resulted from their insensibility *to* a rough elemental life and coarse diet: "inhabitants of a rude and uncultivated climate [were insensible when compared to] those of a polished and civilised nation."[30] The historian of sensibility Barker-Benfield captured such allusions to the powers of external stimuli—an "easiness of being acted upon by external Objects"—as those that "betokened both social and moral status" and depended on climate as a sign of spatial positioning. The body possessed a "memory of place" that converted the physical into the physiological. What filtered into physiology became physiological.[31]

The medicalization of airs lent credibility to the "atmosfear" felt in particular by fragile and ailing constitutions. As Doctor George Cheyne had argued early in the eighteenth century, the greater the sensibility of the nerves, the finer was the constitution, and the larger the (perceived) hazard to those who possessed the faculty. The wounded and women felt and forecast the weather by means of arthritic pain.[32] The robust handled almost any change, but "the infirm cannot well bear the change, and therefore ought to be more careful as to these matters."[33] Physicians increasingly recorded an unusually honed vulnerability of their better-off clientele and their lower threshold of resistance to the *ingesta*: "although there are some individuals capable of bearing great changes of the air, there are others in whom the slightest alteration with respect to density has considerable effect."[34] The extremes of this dependency inspired the literature on "human barometers"—those especially delicate individuals whose bodies "registered" the weather and whose health *was* a weather register. Their health and mood were imagined to mimic the motion of mercury in the barometer, reacting to changes in the ambient medium: "if cold or heat prevail to great excess/more than we ought, we than perspire or less/our passive body alterations finds/and with our bodies sympathize our minds."[35]

But information about the nosopoetic powers of air entered into a wider social ambit. This happened because the physiological studies flagged a problem that transcended the technicalities of animal economy and disease causation. They made it clear that air-dependency defined a body (and mind) which could no longer be treated as subject to internal humors only: an entity molded by forces outside the individual remit of sovereignty, and, by consequence, in need of a new authority in matters of diagnosis, prognosis and therapy. A later writer would rephrase this by saying that "the phenomena of life are not the result of *original organization* only; but that the moral, intellectual, and physical capacities of man are subject to the influences of those causes, the aggregate of which constitutes climate."[36] If climate underwrote national character and political institutions, how much more strongly did it shape individuals and their medico-moral histories? For some, this malleability implied an end of self-sufficiency in the ethics of responsibility.[37] The humorist Thomas Gordon noted a "remarkable Sympathy between a human Spirit and the Weather"; he raised the issue of the principles of moral action resulting from an externally driven physiology that challenged conventional ethics.[38] He observed that as moral decisions and psychological moods could now be seen as no one's *inner* and *individual* accountability—at least not to the extent imagined by the traditional "monadic" selves—then human behavior must be considered to evolve under a diktat from the outside. As the actions of "human barometers" happened without consent and will, the mind was no longer in charge of voluntary decisions and other higher faculties: "it would be very grateing to a man, to hear that the last *Two Thousands* he gave to a *Church* or an *Hospital*, did not follow from an habitual Goodness; but to his Walking up *Constitution Hill* at Seven in the morning without his Breakfast." The "powers without" informed the Enlightenment "aerial body" as they raised objections to the ethics of personal accountability and the atomic self.[39]

THE AIRS OF LUXURY, THE LUXURY OF AIRS

To the ascendant "powers without," the powers "within" responded with alacrity. The astute asked if a society should forfeit its freedom to brute matter in the way the asthmatic would forfeit her health to humidity. David Hume, Lord Kames, and James Dunbar were among the leading anti-climatologists in political theory, but their opposition had to do with the in-sufficiency of physical (as opposed to moral) causes to account for the differences in national characters.[40] More pertinent were the feelings of those like Samuel Johnson who, being something of a pundit in matters of weather talk, decided that, "to call upon the Sun for peace and gaiety, or dep-recate the Clouds lest sorrow should overwhelm us, is the cowardice of Idle-ness, and the idolatry of Folly."[41] Weather moods "operated on luxury" and man ought to shun "the tyranny of the climate, and refuse to enslave his virtue or his reason to the most variable of all variations, the changes of the weather." Johnson's plea only confirmed Cheyne's belief that weather complaints were an exclusive prerogative of cantankerous society types who rationalized the effects of conspicuous consumption by a vulnerability to forces over which they could have no control, from which they could not hide, and which were the causes of whatever blameworthy actions they might otherwise be held responsible for. It is in this sense that the invoca-tion of "environmental" influences in eighteenth-century medicine owed something to the process described by Odo Marquard as "flight into the un-indictability."[42]

The pathology of luxury became a favored theme by the third quarter of the century. Commentators insisted on the shifts in health patterns occa-sioned by the capitalist modes of production and the attendant lifestyle, as Roy Porter explained in the case of the British upper classes. In the criticism of the artificial/abnormal nature of new spaces of lifework, the accent was on urban consumerism, sedentariness, indolence, and fashion, all claimed to impair the stamina of the spending classes. Medical writers from the mid-century sermonized on the evils of opulence; their views appealed to those brought up to treasure rural retreat and agriculture, the ideals dear to British physiocrats. During the 1770s, the view was further pressed forward by a didactic literature that attacked urbanization, the cash nexus, and the bub-ble economy. Landscape poets unearthed the first book of Virgil's *Georgics*, and extolled rural work and pastoralist idyll. They admired simplicity of manners and the salubrity of country air. The peasant archetype became the medical norm against whom the unnatural urbanite was measured in both moral and organic terms.[43]

The medical message behind this utopia was clear: labor made men robust, luxury softened them into disease. Outdoors healed, indoors hurt. When such a dichotomy undergirded the teaching of someone as eloquent and authoritative as Edinburgh's professor of medicine, James Gregory, its glibness must have enthused the students who straddled their provincial origins and urban aspirations between spells of extramural drinking. Gregory reasoned that the men who exposed themselves to the external hardships of the seasons and who were used to tough work, enjoyed vigor like that of healthy animals. "Nor was it till he discovered and brought to perfection artificial means of defending the body generally from the extremities of the seasons, that man was subdued by the severity of his native clime." Those neglectful of shelter "seldom suffered any inconvenience from [the harsh weather]; and it is only in proportion as men become *afraid of exposing* themselves to the intemperance of the air, and sedulously endeavor to avoid it—that they are rendered susceptible of its effects." For Gregory, such calculated shelter was luxury's child, bred by thermometric equability and lethargic relaxation during the winter months. Lowered immunity and slothful loitering were two sides of the same coin.[44]

Medico-moral diatribes against luxury revealed the workings of bourgeois materiality refracted in the thermometric pathology of party-going society. Critics hinted that there was a progressive vulnerability to the exteriors and discovered its sources in social trends that dictated sedentariness, private gathering, and overindulgence. This became everybody's conviction by the last quarter of the century. The renowned physician William Falconer rebuked indoor life so much as to prescribe field labor as a therapy.[45] Dr. Johnson's friend and physician Percival Stockdale averred that the opulent could not endure the activity needed to "preserve the health of the animal machine."[46] Health guides like James Mackenzie and John Armstrong advised activity in the cold to guard against the "drooping sky and indolence,"[47] as comfort amplified sensibility through languor and artificial heating, the new medical pariah. Heat oppressed the skin's nerves and "exposed their extremities" to further stimulation, while sweating rendered cuticles fitter for the transmission of stimuli that resulted in putrefaction. The effect was tragic: a weak, sickly, and obese body nurtured a tropical psyche, at once passionate and amorous, vindictive and timid.[48]

The pathologization of luxury became one of the driving forces behind Enlightened medical environmentalism. The role of the indoors was central, yet it has so far made only a cameo appearance in medical history.[49] The indoors was important for a number of reasons: the time spent in domestic, recreational, and work space in Georgian England could accumulate to three fourths of a life, a good part of which took place away from fresh air. Industrialization added to this plight by lengthening working hours.[50] Houses routinely had fireplaces, wick-lamp illumination, and were brimming with cooking smells that mixed with bodily odors and the vapors from drying clothes. Drafts and bad isolation allowed random circulation, but the air's admission was unintended or a subject of taboo. The night air was especially dreaded for its miasmatic impregnation, a cult that gave a headache or two to those in sealed bedchambers, in between bed baldaquins or impenetrable muslin hangings. It seemed obvious that whatever ailments one acquired during one's life could well be associated with the quality of indoor life. As a German architectural historian remarked, in England, "there is no temptation to spend time out of doors as there is in southern countries; nature does not seduce them into idleness in the open air."[51]

The interest in indoor climates coincided with the associational culture of the public sphere and the urban renaissance. In Europe and America, the new participatory spaces included clubs, lecture theatres, coffeehouses, boudoirs, card rooms, libraries, and scientific meeting places, all of which, being non-purpose built, meant overcrowding, heating, candle smoke, and stifling air. But while historical scholarship has dwelled on how these spaces shaped the intellectual content of the Enlightenment, little attention has been given to the ways in which these spaces affected the *bodies* of the new public.[52] Habermasian public spaces and Foucaldian heterotopias could have been very smelly places. Contemporaries, however, showed more sense. They noted truly "breath-taking" achievements of the Enlightenment including claustrophobia, fainting, overheating, fatigue, and depression of spirits resulting from congestion. Authorities recognized the lethal importance of overcrowding in prisons, workhouses, hospitals, and ships, where contagious fevers decimated unsuspecting inmates. The medical profession launched a total war on the indoors, first with regard to overcrowding, and then, as an epitome of unhealthy living *per se*.[53]

The work of James Makittrick Adair (1728–1802) illustrates these concerns with particular clarity. A member of the Royal Medical Society and the Royal College of Physicians of Edinburgh, early in his career he served as the Commander in chief of the Leeward Islands. His practice took him to Antigua, Andover, and Guildford, during which time he composed a dissertation on yellow fever and the well-received *Natural History of Body and*

Mind (1787). He then moved to Bath to become an expert on resort regimen and fashionable diseases, supporting his earnings with didactic publications like the *Medical Cautions* (1786) and *An Essay on Regimen for the Preservation of Health*. His practice allowed him to observe the habits of the well-off clientele and he paid special attention to residential microclimates. On more than one occasion he railed against the suffocating bedchambers, cardrooms, and parlors where the combined effects of closed windows, fireplaces, tallow candles, rebreathing, and body heat kept the average temperature well above his 58°F recommended standard. Dancing parties and card routs he considered ruinous, chastising his female readership not to attribute their jaundiced-colored eyes to the excess of bile but to the real cause, that is, the tainted air at the card table in a small crowded room where delicate patients often paled like ghosts and blacked out in "swoons," only to catch colds when sobering up on a blustery veranda.[54] (Verandas were the Regency rage that, according to William Cowper, had little rationale in London's sullen seasons). Adair was appalled by the social proxemics that governed such parties: why was it that despite the spaciousness of Bath's public rooms people preferred to cram into bed-chambers, closets, and even cupboards, "not only to the injury of the public institutions, but manifestly, to their own health?"[55] Dense assemblies made sense in London, where spacious apartments allowed breathing. But in Bath's miniscule flats, Adair thought, routs and parties had no medical justification: they thwarted recovery even if they helped physicians' earnings. Adair wrote: "On declaring a few days later, to one of my brethren, a man of humour, my resolution of writing a bitter Phillipic against routs, he archly replied: "let them alone doctor, how otherwise should 26 physicians subsist in this place."[56]

Adair's Phillipics testified to the then-current architectural trends to proliferate the number of private spaces and make a patchwork of house layouts. These, quite literally, resulted in the shrinking of the rooms' air volumes. The mid-eighteenth century rooms of the wealthier people were increasingly cluttered, papered, carpeted, and with windows whose curtains disallowed air flow and facilitated the absorption of odors and dust. The historian Carole Shammas found such changes in "the physical environment of the middle class family [to be] immense."[57] When the compartmentalization of space combined with serious partying and lavish illumination, the air fell as the first victim. Lady Cowper's party in 1769 was lit up by five dozen wax lights, the Duke and Duchess of Northumberland's gala boasted four glass chandeliers each with 25 beeswax candles. In 1798, *The Times* reported of Mr. Thomas Hope's "splendid rout" with one thousand guests roaming

FIGURE 1   Thomas Rowlandson's drawing of a card game illustrating Adair's comments on bodily proximity and closeness of space in late Georgian Britain. From John Steel (ed.), *Mr Rowlandson's England* (London: Antique Collector's Club, 1985), 71.

through 16 rooms (averaging over 60 people per room!), the premises illuminated by 250 wax candles. Modern estimates suggest that the candles alone would make one quarter of the room's air unfit for respiration in under an hour.[58] Priestley was perhaps *not* carried away when, after sniffing from a flask of de-phlogisticated air, he felt that "this pure air may become a fashionable article in luxury (although) Hitherto only two mice and myself have had the privilege of breathing it."[59]

But even as the fashionable diseases such as fainting and hypersensitivity to drafts plagued only a small (if growing) society of people—"valetudinarians in reputation as well as constitution [who] avoid the least breath of air"—physicians did bring to attention the fact that their lifestyles generated a hitherto unknown sensibility to atmospheric agency.[60] One element in this process did raise more concern than others: the sudden change of air. From the mid-eighteenth century, writers on matters of hygiene, regimen, prolongation of life, and general rules of conduct commented on the hazard of exposure to unexpected heats, colds, damps, and drafts found in and around the domestic orbit of life. They warned that thermal shocks compromised health, especially of those who took isothermia for granted. For there was a cold hallway, a damp cellar, an overheated room, a smoky bedchamber, moving between which was to be done only with proper care and attire. The letter of the great Irish political hostess, Emily Mary Fitzgerald, Lady Kildare, to her husband on 9 December 1762, while not quite representative of an average household, captured the predicament well: "I have been here these two days and would you believe it? Starved with cold after coming from Castletown, which shows the coldest houses when constantly lived in will be warmer than the warmest when at any time uninhabited, as is the case here . . . You will say, what, was the print room cold? No, but the way from it from the apartment we are in at present perishingly so—those stairs running with wet, as is the passage above and most of the rooms to this back side of the house; which shows, my love, the necessity of having very, very often fires almost all over the house."[61]

This was in all probability a legitimate complaint. But doctors occasionally dealt with cases in which sensitivity to cold, especially among the sedentary, was breaking records in medical annals. One such person wrote a report to Bristol's Doctor Thomas Beddoes, a political radical and revolutionary medical pneumat, in which he explained that: "what I have to complain of is extreme suffering from cold. [Another doctor] wishes me to increase my quantity of cloathing; I have strongly resisted the suggestion from conviction of its inefficacy. I am wrapped in fleecy hosiery, have thick understockings and silk above. I have, besides, fleecy hosiery drawers, two peticoats, and my gown. I wear long sleeves with gloves under, and a very thick double muslin handkerchief. On going out I put on a cloth pelisse, trimmed with fur close up to the throat, a muff, and a pair of very strong thick shoes over my others. In the house I usually wear warm list shoes also over my others. I have been cloathed in this manner for the last two years, and my chilliness is certainly increased. When I am sitting in a room, I often feel the cold occasioned by opening of a door, as plainly and as immediately in my bowels as I do on my face—I wish to know whether you think more clothing can be beneficial."[62] For the more desperate cases, technology came to the rescue in the form of a steam bath.

Leaving and entering a house became a medical matter. Such actions were considered riskier than coping in a wintry shower because once outside, one was meant to face the adversity. Getting in was a different matter: it was subject to planning. Those who spent time out but ignored the hazards of getting in, were in danger of abusing their constitution already wearied by gambling, dancing, and punch. On frosty days, everybody was advised to acclimatize in unheated antechambers. Physicians disapproved of the skimpy ladies' dress that was no good for nippy hallways, nor for rides in damp carriages. In such dress they were "instantaneously chilled by the sudden transition from suffocating heat to piercing cold."[63] Chills caused internal regurgitation, checked perspiration and eventually led to inflammation, haemorrhages, fevers, and pleurises. But in all this panicky advice press, it was important to stress that the hazards of thermal transitions were hardly of any concern, if they were at all known, to men and women out in the country; in James Johnson's phraseology, it was the delicate female, the pale mechanic, and the sedentary artist who were at risk. For those, urban life and its milieu were "being conceptualized in medical terms."[64]

A METEOROLOGY OF LIFESTYLES

If before 1750 such conceptualization had a quality of mishap and randomness, the last decades of the century saw it in epidemic proportions. By this time, the debated vulnerability of the elites was sometimes discussed in the context of mesmerism, somnambulism, and clairvoyance, which offered an image of bodies that were akin to "marionettes under hidden power."[65] Anthony Florian Madinger Willich, at one time physician to the Saxon Ambassador in London, wrote in 1799 that, next to gout, the "still more general malady of the times, is an extreme sensibility to every change of the atmosphere; or, rather, constantly sensible relation to its influence." Willich noted the delicacy of some of his patients that enabled them to identify the direction of the wind even inside their apartments. He was struck by the "talent so peculiar to our age" which, he thought, was acquired in a most capricious of European climates, and which defined English health as "dependent, frail and transitory."[66]

Was the delicacy in question a property entirely new or was it now only being recognized as a medical issue? Was it a matter of physiology or history? From what we know about the hypochondriacal culture of the early romantic era, it is most likely that the weather susceptibility was a social asset, not somatic liability. The power of air to affect physiology was in some measure at least a response to a demand created by those who, for the lack of a better phrase, *wished* to be affected. This was a familiar theme in medical literature. The Enlightened wits spilled plenty of ink to lambaste those "fond of catching" the English Malady; Adair knew about the patients and the doctors who *chose* a disease, not just a therapy.[67] With the hypersensitive body defined as culture-determined and class-based, the indoor ills could not but become an achievement that marked a select few: the "affluence and luxurious indulgence expose us to distresses" of a kind not suffered by the lower class of society.[68] This polarization not only defined exposure in relation to the wealthy and oversensitized, but—negatively at least—with respect to the poor and robust as well. The sensitivity to air that informed the early nineteenth-century interest in weather as an outdoor phenomenon was in some important elements defined by its indoor- and body-oriented origins. It reflected the medical implications of social and geographic placements that defined the vulnerability of the self-styled modern man.

It would, however, be inaccurate to suggest that the only medical environmentalism known to the early nineteenth century was that limited to the confined, sedentary and well-to-do. In reality, complementing this group's and their doctors' concerns was the often philanthropic concern for the habitations of the poor, urban and rural alike. The concern here was not delimited to the indoors only. Topographic investigation, medical police, and early urban sanitarianism—mostly on paper, though—focused on the outside. The perception of what mattered in the management of the health of the large number of people moved beyond the miniscule space of the card room or dance hall and into the larger if still delimited space of the street, neighborhood, town, and region. It looks as if the outward impetus in medico-moral concern over the quality of nosopoetic living lay in an onion-like expansion that eventually encompassed the totality of what is now, perhaps too indiscriminately, called the "natural environment." If more studies were available on the subject, we might learn whether the driving force in this centrifugal process was the medical ethos that relied on the subtler nervous sensibility and the stronger cultural clout that the hegemonic middle classes used as a lever in correcting the perceived sin that was modern civilization.[69]

Indeed, the middle- and upper-class origins of "environmental" concerns were unquestioned in the period publications. For example, the common view argued that filth was a result of the indifference and ignorance of those who didn't have time and means to share in the genteel sensibility nor be affected by the grumpy old aesthetes revolted by the commoners' dirt. Buttressing this conviction were the accounts from across Britain and the Continent that commented on what was thought to be an acquiescence to a disgraceful state of hygienic affairs in the dwellings of the poor.[70] A Victorian historian of ventilation spoke of the "inveterate public insensibility to pollution, even after the way was shown how to enjoy the blessing of viatic purity." Count Rumford noted the difficulties in prevailing "on the public to accept the boon of improvement even in the matters which come home to every man's business and bosom." The poor were blamed for their ingrained reluctance to give up on unsanitary habits: the municipal acts passed to enforce cleaning of the early modern towns were seen as "acts of rigour and oppression, tending to sacrifice the privileges and enjoyments of the poor to the squeamish feelings and effeminacy of the rich."[71]

What comes out is the argument that the middle- and upper-class noso-poetics of air—together with its concomitant environmentalist outlook—emerged as an ideology that sought to redress the health issues of the population perceived to be bereft of the "sixth" meteorological and sanitary sense. A lack of this sense was blamed as the cause of these people's inability to understand just how densely polluted was the world of their fragile superiors. The poor were rough, insensitive and subversive in their refusal to clean up their environmental act, too thick-skinned to feel the sudden exposures to fireplace drafts. They were too healthy for fresh air and too much outside to care about the barometer. In some cases, their blindness to the weather changes and their obliviousness to muck were on the verge of being a statement of disobedience to the polite culture of weather diarizing and ethic cleansing from above. But what was important to stress in this dichotomous context is a continuity in the nature of concern that spread across the social and spatial scales. In this sense, the early nineteenth-century medics constantly emphasised the *artificiality* of the newly created spaces of work, sleep, and the bacchanal. In their view, these spaces deteriorated not only pathologically—measured in statistics, sights, and smells—but primarily because they transgressed the normal contours of morbidity found in the places governed by divine dispensation of health and disease.

Excess was the key word, debauchery the culprit. Even in Manchester, where the industrial slum caught the eye of the Lit and Phil philanthropists, Richard Price wrote in 1775 that the city suffered a high mortality due to "first, the luxury and the irregular modes of life which prevail in towns; and secondly, the foulness of air."[72] In 1776, a manual advised that fever patients should be kept neither in "pent up close confined nurseries, or small rooms, where the air must be stagnated; nor brought up in great cities or large towns where the air is always polluted by the exhalation of dead animals, or the unwholesome vapours proceeding from standing waters."[73] Furthermore, contemporaries claimed that industry bred hazards that were difficult to understand much less attenuate. While winds, rains, and smog plagued health outside the factory, their action was not as insidious as the factors found in its interiors and its backyards. That the weather came second to man-made conditions was recognized by the same observers who would have otherwise considered urban diseases etiologically commensurable to rural (and more weather-caused) maladies. They were now, however, faced with a higher risk that implied new conditions and required a qualitatively different cure. John Walker, the medical topographer of Huddersfield, wrote of the need to address what he called the "*anomalous diseases* perpetually presenting themselves among the manufacturing population," arising from dirt, overcrowding, and bad air.[74]

In the same decade, the Aberdeen-educated physician Colin Chisholm, in discussing typhus, conjectured that it was possible that "the *unnatural* state in which the inhabitants of manufacturing towns are placed," may dispose them to the disease. Unlike the manufacturing districts, the commercial towns' bustle preserved the balance of the inhabitants. Commerce was invigorating, but excessive industry, for which the human body was *not intended* and into which it was steeped by artificial wants, caprice, or human avarice, made the organism retentive of poisons and prone to infection. Typhus was a retribution for the overstretching of the natural powers in places which had carved their own "un-natural" regimes and required new programs of remedies.[75] James Johnson, perhaps the highest contemporary authority on medical meteorology (with James Clark), thought that intellectuals, bon-vivants, and businessmen thwarted their naturally endowed energies by exposure to civilized living that was steeped in stress, deadlines, partying, and the lack of outdoor activity. "Especially in this country," Johnson observed, "and in consequence of the immense interest in politics, religion, commerce, literature, [and] a stress due to speculative risks [. . .] it is astonishing to observe the deleterious influences of the mental perturbations on the functions of the corporeal fabric."[76] In the centres of commerce and fashion, the "natural" weather had very little medical relevance. Instead, the mapping of early (industrial and urban) space reflected a medico-moral criticism of industrial extravaganza that was presented as the sole reason for the physical deterioration of both its victims and its consumers.

It seems reasonable to suppose that the combination of medical and moral analyses of atmospheric exposures heralded a crucial change in the history of European Hippocratism and medical meteorology. As the discussion on indoors and workspaces indicated, a relatively contained problematic aimed to address the conditions of a small number of people grew by the early nineteenth century into a more general discussion of the pathological effects of physiological, atmospheric, and social agencies subsumed under the notions of sensibility, vulnerability, and differential placements. The medical blight of modern Europeans was not only in what they could, but failed, to rectify (*bad* air), but in what they could never hope to escape (*air* itself). "Air was the one of non-natural affecting the bodies every moment of our lives and from which there was not escape, for good or ill."[77] The resultant "airmindedness" introduced a patient whose condition depended on contexts, not an immediate pathological change: a person was considered ill not because of humoral imbalance, but because he or she had problems coping with heat.[78]

In identifying the physiology of air as the main source of the medico-moral airmindedness of the late Enlightenment, and in seeing the indoor and urban lifestyles as the target of medical meteorology, this essay tried to identify the factors behind the distinction between the body and its meteorological surroundings on the eve of modern meteorological thinking. The "outside" weather as a medical entity was only marginal compared to the mainstream interest in the pathology of small-space conditions. This means that sensibilities to environmental exposure were not expressions of some die-hard Hippocratism, but an epitaph of the changes in lifework that brought about a new vulnerability expressed with regard to bodies in space, seasons, and society. The susceptibility to atmospheric agency can thus be seen as reifying—perhaps fetishizing—the anxiety over exterior risks in that it worked under the assumption of health as a commodity under a threat. But as the literature on preservation of health and prolongation of life testified, this was also a commodity that could be maintained, augmented, or lost to the misrule of life. Paraphrasing the medical historian William Coleman, the threat from the outside provided a "framework for articulating the primary demands imposed by the conditions of existence upon men and women who sought seriously to preserve their physical well being." With this in mind, we may begin to conceive how the medico-moral concerns made inroads into a physiology of differential exposures that, consequently, informed the limits of a medical meteorology as an expert construction of mundane weather.[79]

What are the implications for the history of meteorology? I'd like to return to the argument made in the introduction that the historical task should involve correlating the lived with the measured weather. Taking the latter as irreducible—because such a reduction would mean subjectivity, error, and non-standardizable judgment—might be justified within an institutionally based observational meteorology as we know it today. But this would be inappropriate in discussing the times when such measurements were about to be institutionalized. To buttress this, let us look at a snapshot from the 1850s debate on the medico-climatological qualities of the Atlantic island of Madeira in curing tuberculosis. The excerpt is from a book written by James Mackenzie Bloxam, a Denbighshire barrister and a regular visitor to the island. The book contains a vitriolic response to the writing of other medical climatologists (John Abraham Mason and Thomas Burgess) in which the latter had used hygrometric data to disqualify the island's curative powers. In his response to the two, Bloxam prided himself on being a layman in matters of climatotherapy precisely because he had seen so much misuse of instrumental observations that it had "created in me a distaste for pursuit of science."

Bloxam challenged the scientific method of those who think "that a few hygrometric and thermometric data are sufficient to enable them to pronounce that climate of long-established reputation, whether good or bad, is in fact the reverse of what was previously supposed." But data, Bloxam argued, were often fictions concocted by badly calibrated instruments and read by people without qualification and experience. Even the terminology was suspect: the epithets that the "scientific" authors used to describe Madeira's weather—such as "dampness" and "dryness"—were arbitrarily defined to be "in accordance with [scientists'] method of estimating those qualities. Is it not obvious that the sensation of dampness depends on something besides the number of grains of water, in a cubic foot of air, and that dr Mason's satisfactory manner of estimating dampness fails to detect that *not unimportant something* whatever that may be?" Bloxam's "not unimportant something" merits much more attention in historical reconstructions of meteorology.[80]

ACKNOWLEDGMENTS

This project has been made possible by the generous support of the Wellcome Trust for the History of Medicine. I would like to thank John Pickstone and Mark Harrison for comments on an early draft. A shorter version of this has been presented before the members of the University of Cambridge's History and Philosophy of Science Department, to whom I owe gratitude for excellent suggestions and hospitality.

NOTES

1. Albert Einstein, "Autobiographical Notes," in P.A. Schipp, ed. *Albert Einstein: Philosopher-Scientist* (Evanston: Library of Living Philosophers, 1949), 5.

2. George Sigmond, "Materia Medica and Therapeutics," *Lancet* 1837 (290–95), 293.

3. Clarence J. Glacken, *Traces on the Rhodian Shore: Nature and Culture in Western Thought from Ancient Times to the End of the Eighteenth Century* (Berkeley: University of California Press, 1976), 604.

4. L. J. Jordanova, "Earth Science and Environmental Medicine: the Synthesis of the Late Enlightenment," in L. J. Jordanova and Roy S. Porter (eds.), *Images of the Earth: Essay in the History of the Environmental Sciences* (BHSH: London, 1978), 119–146. The terminology is discussed in James Riley, *The Eighteenth-Century Campaign to Avoid Disease* (New York: St. Martin Press, 1987), xv–xvi. David Arnold, *Colonizing the Body: State Medicine and Epidemic Disease in Nineteenth-Century India* (Berkeley: University of California Press, 1993). For an authoritative treatment of the colonial context, see Mark Harrison, *Climates and Constitutions: Health, Race, Environment and British Imperialism in India* (Oxford: Oxford University Press, 1999).

5. Hans-Joachim Voth, "Time Use in Eighteenth-Century London: Some Evidence from the Old Bailey," *Journal of Economic History* 57, No. 2 (1997): 497–499; Caroline Hannaway, "The Societe Royale de Medecine and Epidemics in the Ancient Regime," *Bulletin of Medical History* 46 (1972): 257–273; Caroline Hannaway, "Medicine, Public Welfare, and the State in Eighteenth Century France: The Société de Medecine de Paris (1776–1793)," Ph.D. Johns Hopkins 1974; Ludmilla Jordanova, "Policing Public Health in France, 1780–1815," in Teizo Ogawa (ed.), *Public Health* (Tokyo: Taniguchi Foundation, 1981), 12–32; Henry Lowood, *Patriotism, Profit and the Promotion of Science in the German Enlightenment* (New York and London: Garland, 1991); Mary Lindemann, *Health and Healing in Eighteenth Century Germany* (Baltimore and London: The John Hopkins University Press, 1996), 271–272.

6. Carlo M. Cipolla, *Miasmas and Disease: Public Health and the Environment in the Pre-industrial Age* (New Haven and London: Yale University Press, 1992); For developments in the Low Countries see Gaston R. Demaree, "The Neo-Hippocratic Hypothesis—An Integrated 18th Century View of Medicine, Climate and Environment," *Zeszyty Naukowe Uniwerytetu Jagiellonskiego* 102 (1996): 515–518. The poisonous airs as "venomous exhalations" are treated in John Prestwich, *Dissertation on Mineral, Animal, and Vegetable Poisons* (London: F. Newberry, 1775), 71–106.

7. Simon Schaffer, "Measuring Virtue: Eudiometry, Enlightenment and Pneumatic Medicine," in Andrew Cunningham and Roger French (eds.), *The Medical Enlightenment of the Eighteenth Century* (Cambridge: Cambridge University Press,

1990), 281–318; on travel and health, David Arnold (ed.), *Warm Climates and Western Medicine: The Emergence of Tropical Medicine, 1500–1900* (Amsterdam: Rodopi, 1996); Christopher Hoolihan, "Health and Travel in Nineteenth-Century Rome," *Journal of the History of Medicine and Allied Sciences* (1989): 462–485; on ventilation and similar measures see James Riley, *The Eighteenth-Century Campaign to Avoid Disease* (New York: St. Martin's Press, 1986); on central heating see Todd Wilmert, "Heating Methods and Their Impact on Soane's Work: Lincoln Inn's Fields and Dulwich Picture Gallery," *Journal of the Society of Architectural Historians* 52 (1993): 26–58; on the changing of house layout and infrastructure, Annik Pardailhe-Galbrun, *The Birth of Intimacy: Privacy and Domestic Life in Early Modern Paris* (Cambridge: Polity Press, 1991); on the umbrella T. S. Crawford, A *History of the Umbrella* (Newton Abbot: David and Charles, 1970); suburban England is the subject of Carl Estabrook, *Urbane and Rural England: Cultural Ties and Social Spheres in the Provinces* (Manchester: Manchester University Press, 1998); on tennis courts and gymnasia Henning Eichberg, "The Enclosure of the Body—On the Historical Relativity of "health," "nature" and the Environment of Sport," *Journal of Contemporary History* 21 (1986): 99–121; on "mephitic" theory and purpose built cemeteries, Thomas W. Laqueur, "The Places of the Dead in Modernity," in Colin Jones and Dror Wahrman (eds.), *The Age of Cultural Revolutions: Britain and France, 1750–1820* (Berkeley: University of California Press, 2002), 17–32.

8. Francis Lobo, "John Haygarth, Smallpox, and Religious Dissent in Eighteenth-century England," in Andrew Cunningham and Roger French (eds.), *The Medical Enlightenment of the Eighteenth Century* (Cambridge: Cambridge University Press, 1990), 217–253; Christopher Lawrence, "Disciplining Disease: Scurvy, the Navy and Imperial Expansion, 1750–1820," in David Phillip Miller and Peter Hans Reill (eds.), *Visions of Empire: Voyages, Botany and Representations of Nature* (Cambridge: Cambridge University Press, 1996), 80–106; early works on disease and tropics include William Hilary, *Observations on the Changes of the Air and Concomitant Epidemical Diseases* (London: L. Hawes, 1766); George Cleghorn's letters to John Fothergill on the diseases on the West Indies (1737–1749) were published as part of his *Observations on the Epidemical Diseases in Minorca* (London: D. Wilson, 1751); J. B. Schotte, M.D., "Journal of the Weather at Senegambia, during the Prevalence of a Very Fatal Putrid Disorder," *Philosophical Transactions* 70 (1780): 478–506.

9. See Philip J. Frankenfeld, "Technological Citizenship: A Normative Framework for Risk Studies," *Science, Technology and Human Values* 17 (1992): 459–484.

10. The architect John Wood remarked of the English cottages "that they were unhealthy from the lowness and closeness of the rooms; from their facing mostly the north and west; and from the chambers being crowded into the roof; where having nothing to defend them from the weather but the rafters and bare roof

without ceiling; they were stifling hot in the summer and freezing cold in the winter [. . .] the dormers leaky added greatly to the dampness, unhealthiness and decay of the cottage." John Wood, *A Series of Plans for cottages or habitations of the labourer* (London: I. and J. Taylor, 1792), 2.

11. For recent discussion on the city/body analogy see Joel A. Tarr, "The Metabolism of the Industrial City," *Journal of Urban History* 28 (2002): 511–545; Sally Sheard and Helen Power, *Body and City: Histories of Urban Public Health* (Aldershot: Ashgate, 2000).

12. In Britain, for example, the main protagonists included Clifton Wintringham, *Commentarium nosologicum morbos epidemicos et aeris variationes in urbe Eboracenci locisque vicinis, ab anno 1715, usque ad finem anni 1725* (Londini: J. Clark, 1727); John Huxham, *Observations of the air and epidemical diseases, made at Plymouth from 1728–1737* (London: J. Hinton, 1759). Charles Bisset, *An essay on medical constitution of Great Britain to which are added observations on the weather* (London, 1762); John Rutty, *A chronological history of the weather and seasons and of the prevailing diseases in Dublin* (London: A. Millar, 1770); Thomas Short, *A General Chronological History of the Air, weather, seasons, meteors etc in sundry places and different times* (London: T. Longman, 1749); John Murray, MD, "Journal containing daily meteorological and monthly medical observations," Wellcome MS.7840. See also "Meteorological Observations [in Bristol, January 1774]," Wellcome Trust Library MS. MSL. 111. See also Andrea Rusnock, "Hipocrates, Bacon, and Medical Meteorology at the Royal Society, 1700–1750," in David Cantor (ed.), *Reinventing Hippocrates* (Aldershot: Ashgate, 2002), 136–153; idem., *Vital Accounts: Quantifying Health and Population in Eighteenth-Century England and France* (Cambridge: Cambridge University Press, 2002). Among the first official weather records in the United States was that inaugurated in 1814 by the Army Medical Department, under General Joseph Lovell, see W. J. Humphreys, "Origin and Growth of the Weather Service of the United States, and Cincinnati's part therein," *Scientific Monthly* 18 (1924): 372–382.

13. J. M. Walker, "The Meteorological Societies of London," *Weather* 48 (1993): 364–372.

14. James Tatem, "Letter to the Editor," *Monthly Magazine* 55 (1823): 207.

15. Hennen, Sketches of the Medical Topography, xvi.

16. For example, in the case of English, "the climate of the North cannot fill up the measure of contentment to its inhabitants; there ever remains a want, and it must be supplied by liberty or by honour, or by profit; it creates a degree of restlessness, which is always seeking for something more." Climate acted on the "fabric of our being" with "powerful energy" an "irresistible impulse" that had implications in the "economy of civil life." *Observations on the power of climate over the policy, strength, and manners, of nations* (London: J. Almon, 1774), 155. See also Sara Warneke, "A Taste for Newfangledness: The Destructive

Potential of Novelty in Early Modern England," *Sixteenth Century Journal* 26 (1995): 881–895. James Dunbar, *Essay on the History of Mankind in Rude and Cultivated Ages* (London: W. Strahan, 1781), 221–222. See also Christopher J. Berry, " 'Climate' in the eighteenth century: James Dunbar and the Scottish case," *Texas Studies in Literature and Language* 16 (1974): 281–292.

17. Andrew Hamper, *The Economy of Health* (London: Author, 1785), 1.

18. An early suggestion that nosopoetics deserved attention is found in John Adams's early nineteenth-century explanation that the disease of *atmospheric* causes were those not caused by the airborne matter—the purest atmosphere could be a carrier—but by the physical properties of the air. "Nothing can be called a contagion unless the person affected by it can induce a similar disease in others *without* regard to season, climate, or any local circumstance." Quoted in Benjamin Ward Richardson, *The Field of Disease. A Book of Preventive Medicine* (London: MacMillan, 1883), 729–730.

19. George Adams, *Lectures on Natural and Experimental Philosophy* (London: R. Hindmarsh, 1794), 12. "The notion that the barometer might indicate changes that the human body was also undergoing was, therefore, an intriguing suggestion but also a cryptic one. The "human barometer" signified a relationship between the body and the atmosphere, of which the conscious mind was not entirely aware, and which seemed in some way to foretell the future." Jan Golinski, "The Human Barometer: Weather Instruments and the Body in Eighteenth-Century England," Paper given at the American Society for Eighteenth-Century Studies Annual Meeting, Notre Dame, Indiana, 3 April 1998, 2.

20. Benjamin Franklin, "A Letter from Dr. Benjamin Franklin to Dr. Ingenhausz, Physician to the Emperor, at Vienna," American Philosophical Society *Transactions* 2: 1–27, quoted in John Crowley, *The Invention of Comfort: Sensibilities and Design in Early Modern Britain and Early America* (Baltimore: The Johns Hopkins University Press, 2001), 187.

21. On skin diseases, see Daniel Turner, *De Morbis Cutaneis*, 1714, 86. On the term nosopoetics, see John Arbuthnot, *An Essay Concerning the Effects of Air on Human Bodies* (London: T. Johnson, 1733), 2. On Arbuthnot's political and scientific affiliations, see David E. Shuttleton, " 'A Modest Examination': John Arbuthnot and the Scottish Newtonians," *British Journal for Eighteenth Century Studies* 18 (1995): 47–62. Among the many quips on the necessity of air see Jeremiah Wanewright, *Mechanical Account of the Non-Naturals* (London: J. Clarke, 1722), 51; Ebenezer Sibly, *The medical mirror. Or treatise on the impregnation of the human female. Shewing the origin of diseases and the principles of life and death*. London, [1800?].

22. James Drake's *Anthropologia Nova* (1707), pp. The move towards iatromechanism is discussed in Theodore Browne, *The Mechanical Philosophy and the "Animal Economy"* (Arno Press: New York, 1981); Anita Guerrini, "Archibald Pitcairne and Newtonian Medicine," *Medical History* 31 (1987): 70–83;

23. Albrecht von Haller's doctrine of nerves provided a rationale as it suggested "the refined sensibility" that, among other things, afforded a medicine concerned with "the placement of the body within the environment." Barbara Duden, *The Woman Beneath the Skin: A Doctor's Patients in Eighteenth Century Germany* (Cambridge, Mass.: Harvard University Press, 1991), 14. Christopher Lawrence, "The Nervous System and Society in the Scottish Enlightenment," in Steven Shapin and Barry Barnes (eds.), *Natural Order: Historical Studies of Scientific Culture* (London: Sage Publications, 1979), 19–40.

24. Roselyne Rey, "Vitalism, Disease and Society" in Roy Porter (ed.), *Medicine in the Enlightenment* (Amsterdam: Rodopi, 1995), 278; Lester King, "Some Problems of Causality in 18th Century Medicine," *Bulletin of the History of Medicine* 37 (1963): 15–24. On the non-naturals see Luis Garcia-Ballester, *Galen and Galenism* (Aldershot: Ashgate Variorum, 2002). "On the Organs of the 'Six Non-natural things' in Galen" (117–152). Jerome J. Bylebyl, "Galen on the Non-Natural Causes of Variations in the Pulse," *Bulletin of the History of Medicine* 45 (1971): 482–845 writes: "Not by nature," in medicine, designated those things (*res non naturales* in medieval Latin terminology) which, though necessary for life, were not part of man's natural endowment." See also *Hippocrates's treatise on the preservation of health. Wherein is explained the salutary and pernicious effects, on different constitutions* (London: J. Bell, 1776).

25. Williams, *The Physical and the Moral*, 153. On hygenists who rendered the "environment [. . .] social in nature," see William Coleman, *Death is a Social Disease: Public Health and Political Economy in Early Industrial France* (Madison: University of Wisconsin Press, 1982), 202.

26. Laurence Brockliss and Colin Jones, *The Medical World of Early Modern France* (Oxford: Clarendon Press, 1997); see also Pedro Entralgo, *Mind and Body: Psychosomatic Pathology: A Short History of the Evolution of Medical Thought* (London: Harvill, 1955).

27. See James Gregory, *A Dissertation on the Influence of Change of Climate in Curing Diseases*; translated from the original Latin and enlarged with occasional Notes by William P. C. Barton, M.D. (Philadelphia: Thomas Dobson, 1815). Medical dissertations on air include Robert Gusthart, *Specimen Medicum Inugurale de Aere ejusque in respiratione usu et effectibus* (Edinburgh: Thomas Ruddimann, 1740); Robert Willan, *Dissertatio medica inauguralis de Qualitatibus Aeris* (Edinburgh: James Cheyne, 1745); Macfait, *Dissertatio Medica Inauguralis de Aere, Aquis et Locis* (Edinburgh: 1745); James Johnstone, *Tentamen Medicum Inaugurale de Aeris factitii imperio in primis corporis humani viis* (ed. T. and W. Ruddimannos, 1750); John Campbell, *Dissertation medica inaugralis, de Aere quatenus mrorborum causa* (Edinburgh: Murray and Cohran, 1754); Daniel Rutherford, *Dissertation inauguralis de aere fixo dicto, aut mephitico, etc.* (Edinburgh: Balfour and Smellie, 1772); Edmundus Cullen, *Tentamen Medicum inugurale de Aere, et Imperio eju in Corpora Humana*

(Edinburgh: Balfour and Smellie, 1781); Henri Burton, *Tentamen Physiologico-Medicum de Usu et Effectu Aeris Puri in Corpus Humanum* (Edinburgh: Balfour and Smellie), 1788; Henry Robertson, *Dissertatio chemica medica inauguralis, de aere atmosphaerico* (Edinburgh: C. Stewart, 1801); William Cheekes, *Disputatio Chemica Inauguralis de Aere* (under George Baird) (Edinburgh: Adam Neill and Co., 1803); Nicholas Pitta, *Dissertatio physiologica inauguralis, de caeli effectu in genus humanum* (Edinburgh: C. Stewart, 1812); George Samuel Jenks, *Dissertatio medica inauguralis de coelo tabescentibus benigno* (Edinburgh: J. Pillans, 1821); William Jackson, *Tentamen chemica medica inauguralis de aëre communi* (Edinburgh: P. Neill, 1822).

28. Thomas Broman, "The Medical Sciences," in Roy Porter, (ed.) *The Cambridge History of Science: vol. 4, Eighteenth-Century Science* (Cambridge: Cambridge University Press, 2003), 463–484. Treatises on the regimen, health preservation and non-naturals rose in popularity; William Buchan's *Domestic Medicine* went through 20 editions by 1797; The University's *Medical and Philosophical Commentaries*, edited during the 1770s by Andrew Duncan, kept the faculty abreast with the Montpellier vitalism.

29. Lawrence, "The Nervous System and Society in the Scottish Enlightenment"; B. White, "Scottish Medicine and the English Public Health," in Derek Dow (ed.), *The Influence of Scottish Medicine* (Carnfort: Pantheon, 1988); George S. Rousseau, *Nervous Acts: Essays on Literature, Culture and Sensibility* (Palgrave: London, 2004).

30. Lawrence, "The Nervous System," 29. Note Cullen's personification of climate as "rude and uncultivated," linking the shaper with the shaped.

31. G. J. Barker-Benfield, *The Culture of Sensibility: Sex and Society in Eighteenth-Century Britain* (Chicago: The Chicago University Press, 1996), 9. See also R. K. French, *Robert Whytt, the Soul, and Medicine* (London: The Wellcome Institute for the History of Medicine, 1969). On the bodily memory see Severine Pilloud and Micheline Louis-Courvoisier, "The Intimate Experience of the Body in the Eighteenth Century: between Interiority and Exteriority," *Medical History* 47 (2003): 451–472.

32. Katharine Anderson, "Instincts and Instruments," in Christopher D. Green, Marlene Shore and Thomas Teo (eds.), *The Transformation of Psychology: Influences of 19th-Century Philosophy, Technology, and Natural Science* (Washington: American Psychological Association, 2001), 165.

33. Huxham, *Observations*, iv.

34. Gregory James, "Theory of Medicine, Notes of Lectures" (Edinburgh, ca. 1785), Wellcome Trust Library MS 2597.

35. Mr. Phelps, *The Human Barometer: or the Living Weather Glass. A Philosophic Poem* (London: M. Cooper, 1743), 15. See also Terry Castle, *The Female Thermometer: Eighteenth-Century Culture and the Invention of the Uncanny* (Oxford: Oxford University Press, 1995), Golinski, "Human Barometer." In *The*

English Malady, Cheyne explained that the English climate, which in Voltaire's view was so dull as to breed misanthropy, the soil, sedentary occupation, the growth of towns and luxury "have brought forth a class of distemper with atrocious and frightful symptoms, scarce known to our ancestors," cited in Oswald Doughty, "The English Malady of the Eighteenth Century," *Review of English Studies* 2 (1926): 257–269, 259.

36. Quotation from a review of Samuel Forry, *The Climate of United States and its Endemic Influences*, by "CC" for *The Western Journal of Medicine and Surgery* 7 (1843): 142–153, 146. My italics.

37. Richard Olson, "The Human Sciences," in Roy Porter (ed.), *The Cambridge History of Science: vol. 4, Eighteenth-Century Science* (Cambridge: Cambridge University Press, 2003), 436–462.

38. On an analogous process of separation of behavior from moral philosophy see Jorge Arditi, "Hegemony and Etiquette: An Exploration on the Transformation of Practice and Power in Eighteenth-Century England," *British Journal of Sociology* 45 (1994): 177–193. See also Malcolm Bull, "Secularization and Medicalization," *British Journal of Sociology* 41 (1990): 245–261.

39. Thomas Gordon, *The Humourist* (London, 1720), 90. The psychic phenomena did not always—nor even for the most part—respond to the "influences to something within ourselves" but to what was external: "the mind rises and falls, quickens or stagnates, just as the Operation of the Powers without direct or relieve it . . . it would be very grateing to a man, to hear that the last *Two Thousands* he gave to a *Church* or an *Hospital*, did not follow from an habitual Goodness; but to his Walking up *Constitution Hill* at Seven in the morning without his Breakfast. It was vanity to allow for such dependency in animals (as in the form of instinct), and not see the humans enslaved by the same force. Apply instinct to a man and "we are affronted: we cannot bear to have our best Thoughts in *Poetry* owing to a *Heath* or a *Hill*, or our Speeches in the House to a *cool Walk in the Garden*." Ibid., 90–91.

40. David Hume, "On National Characters," in *Essays and Treatises on Several Subjects*. (London: A. Millar, 1758), 128–137. See also Christopher Berry, "Climate in the Eighteenth Century: James Dunbar and the Scottish Case," *Texas Studies in Literature and Language* 16 (1974): 281–292.

41. *The Idler by the Author of the Rambler* (London: J. Rivington, 1790), 50.

42. Odo Marquard, *Farewell to the Matters of Principle* (Oxford: Oxford University Press, 1989).

43. This ideology coincided with the literal appearance of idyllic cottages: commissioned by urban clientele, the architectural celebrities Repton, Plaw, and Soane, added comfort and the picturesque onto the simplicity of rural dwellings that, although criticized for stylistic affection, appealed to educated taste and enjoyed success through a combination of exoticism and authentic "Englishness." These developments coincided with the appeal of simplicity and understatement that

29

shaped the neo-classical informality of "undress" based on an explicitly medical reasoning. See Crowley, *The Invention of Comfort*, 224, and Geoffrey Squire, *Dress, Art and Society 1560–1970* (New York, 1974), 103–116. On the literary cultures of physiocracy and rural retreat, see Vladimir Jankovic, "Arcadian Instincts: A Geography of Truth in Georgian England," in Miles Ogborn and Charles Withers (eds.), *Georgian Geographies: Space, Place and Landscape in the Eighteenth Century* (Manchester: University of Manchester Press, 2003), 174–191.

44. Gregory, *A Dissertation on the Influence of Change of Climate in Curing Diseases*, 30. The French context is elaborated in Anne C. Vila, "Beyond Sympathy: Vapours, Melancholia, and the Pathologies of Sensibility," *Yale French Studies* 92 (1997): 88–101.

45. William Falconer, *An essay on the preservation of the health of persons employed in agriculture, and on the cure of the diseases incident to that way of life* (London: R. Cruttwell, 1789).

46. Percival Stockdale, *Three Discourses: Two against Luxury and Dissipation. One on universal benevolence* (London: W. Flexney, 1773), 8.

47. John Armstrong, *The Art of Health*, etc; James Mackenzie, *The history of health, and the art of preserving it* (Edinburgh, 1758).

48. William Falconer, *Remarks on the Influence of Climate* (London: C. Dilly, 1781), 508, 4.

49. For the "outdoor" focus of the public health movement, see Anthony Wohl's *The Eternal Slum: Housing and Social Policy in Victorian London* (Edward Arnold: London, 1977), 2–14. But see Paul Langford, *Englishness Identified* (Oxford, 2000).

50. Hans-Joachim Voth, "Time and Work in Eighteenth-Century London," *The Journal of Economic History* 58 (1998), pp. 29–58.

51. Herman Muthesius, *The English House* (London: Crosby Lockwood Staples, (1904) 1979), 8.

52. But see Alain Corbin, *The Foul and the Fragrant* (Leamington Spa: Berg, 1986). The literature on the public sphere is extensive and growing. See Margaret Jacob, "The Mental Landscapes of the Public Sphere," *Eighteenth Century Studies* (1994): 95–113.

53. Robin Evans, *The Fabrication of Virtue: English Prison Architecture, 1750–1840* (Cambridge: Cambridge University Press, 1982).

54. James Makittrick Adair, *Medical Cautions, for the Consideration of Invalids; those Especially who Resort to Bath: Containing Essays on Fashionable Diseases* (Bath: R. Cruttwell, 1786), 43.

55. Ibid., 35. Could such practice be an example of a "trickle-up" effect among the elites learning about the erotic conveniences of small-spaces from their servants? Tim Meldrum, "Domestic Service, Privacy and the Eighteenth-century Metropolitan Household," *Urban History* 26 (1999): 27–39, 38. This is a distinct

possibility given the fresh-air proselytizing. William Buchan for example advised mothers to take children outdoors, as the closed atmospheres weakened the constitution. He compared indoor children to greenhouse plants not destined for "the strength, vigour and magnitude" of their outdoor counterparts. William Buchan, *Domestic Medicine: or, a treatise on the prevention and cure of diseases by regimen and simple medicines* (London: W. Strachan, 1772), 39.

56. Adair, *Medical Cautions*, 44.

57. Carole Shammas, "The Domestic Environment in Early Modern England and America," *Journal of Social History* 14 (1980): 4–24; Lawrence Stone, "The Private and the Public in the Stately Homes in England, 1500–1990," *Social Research* 58 (1991): 227–257, 256; Frank E Brown, "Continuity and Change in the Urban House: Developments in Domestic Space Organisation in Seventeenth-Century London," *Comparative Studies in Society and History* 28 (1986): 558–590; Tamara Hareven, "The Home and the Family in Historical Perspective," *Social Research* 58 (1991): 253-286. On innovation in domestic comfort across Europe see Rafaella Sarti, *Europe at Home: Family and Material Culture 1500–1800* (New Haven: Yale University Press, 2002), 96ff.

58. *The Times*, May 10, 1802, 2. A good-sized room (20 ft times 13 ft times 10 ft high) would contain 200 pounds weight of air. A single pound of charcoal (equivalent to roughly 10 candles) would burn a quantity of oxygen contained in 14 pounds of atmospheric air. Tomlison, who made this calculation, thought that this meant 50 lbs of air unfit for respiration and the need for the air renewal "many times an hour." Charles Tomlison, *Rudimentary Treatise on Warming and Ventilation* (London: 1850), 52. Some earlier works reflecting on the building practices from a meteorological perspective include William Halfpenny, *Useful architecture in twenty-one new designs for erecting parsonage-houses, farm-houses, and inns; with their respective offices* (London, 1752); William Pain, *The builder's companion, and workman's general assistant* (London, 1769); Isaac Ware, *A complete body of architecture. Adorned with plans and elevations, from original designs* (London, 1768).

59. Joseph Priestley, *Experiments and observations on different kinds of air, and other branches of natural philosophy, connected with the subject* (Birmingham, 1790).

60. Richard Brinsley Sheridan, *The school for scandal, a comedy, as it is performed at the Theatre Royal, Drury-Lane* (London: E. Powell, 1798), 14.

61. Quoted in John Fowler and John Cornforth, *English Decoration in the 18th Century* (London: Barrie and Jenkins, 1974), 225.

62. Beddoes, *Manual Of Health*, 224. On Beddoes, see Roy Porter, *Doctor of Society: Thomas Beddoes and the sick trade in late-Enlightenment England* (Routledge: London, 1992). Little is written on medicine and clothing in the European lands at the turn of nineteenth century when the topic flooded the advice literature, for example, Buchan, *Domestic Medicine*, Walter Vaughan, *An Essay,*

Philosophical and Medical, Concerning Modern Clothing [Rochester], 1792. See also Edward Tobias Renbourn, *Materials and clothing in health and disease: history, physiology and hygiene: medical and psychological aspects* (London: K. K. Lewis, 1972).

63. Adair, *Medical Cautions*, 36.

64. On the interest in sudden changes of air see Jones and Brockliss, *The Medical World of Early Modern France* (Oxford: Clarendon Press, 1997), 463. On medicalization of everyday life see Barbara Duden, *The Woman Beneath the Skin: A Doctor's Patients in Eighteenth Century Germany* (Harvard University Press: Cambridge, Mass., 1991), 141, where she shows how "the hierarchy of illness-causing phenomena was conceived as part of the logic of a life story, not as a part of the logic of the body as such." See also Jordanova, "Earth Sciences and Environmental Medicine," 121. Jan Golinski surmises that the emphasis on rapidity of change followed by the widespread domestic use of the barometer, which emphasized short-term fluctuations, rather than the longer weather patterns. Golinski, "The Human Barometer," 2.

65. Simon Schaffer, "Enlightened Automata," in Clark, Golinski and Schaffer (eds.), *The Sciences in Enlightened Europe*, 158. See also Mary Terrall, "Metaphysics, Mathematics and the Gendering of Science in Eighteenth Century France," in ibid., 246–271.

66. A. F. M. Willich, *Lectures on diet and regimen: being a systematic inquiry into the most rational means of preserving health and prolonging life* (London: A. Strahan, for T. N. Longman and O. Rees, 1800), 58–59. Nor was this a peculiarly English phenomenon.

67. Jeremy Collier, *Essays upon Several Moral Subjects* (London: Richard Sare, 1702–3), 42; for an aestheticization of consumption and its social cachet and value see Clark Lawlor and Akihito Suzuki, "The Disease of the Self: Representing Consumption, 1700–1830," *Bulletin of the History of Medicine* 74 (2000): 458–494; George S. Rousseau and Roy Porter, *Gout: the Patrician Malady* (New Haven: Yale University Press, 1998).

68. Adair, *Medical Cautions*, 53.

69. For a discussion of the contingencies involved in the constitution of urban space see Mark Jenner, "Underground, Overground: Pollution and Place in Urban History," *Journal of Urban History* 24 (1997): 97–110. See also Christine Mesiner Rosen and Joel Arthur Tarr, "The Importance of an Urban Perspective in Environmental History," *Journal of Urban History* 20 (1994): 299–310.

70. For the living conditions of European peasantry, see Jerome Blum, *The End of The Old Order in Rural Europe* (Princeton: Princeton University Press, 1978), 178–183; also Alexander Fenton, *The Hearth in Scotland* (Dundee: National Museum of Antiquities, 1981). On timber hardening, Peter Brimblecombe, "Interest in Air Pollution among Early Fellows of the Royal Society," *Notes and Records of the Royal Society* 32 (1978): 123–129; John Fitchen, "The Problem

of Ventilation through the Ages," *Technology and Culture* 22 (1981): 485–511. The nineteenth-century campaigns against indoor pollution are treated by Stephen Mosley, "Fresh Air and Foul: the Role of the Open Fireplace in Ventilating the British Home, 1837–1910," *Planning Perspectives* 18 (2003): 1–21.

71. Walter Bernan, *On the History and Art of Warming and Ventilating Rooms and Buildings* (London: George Bell, 1845), 90, iv. Better "environments" had apparently very little to offer to the European underclasses. This aspect of grass root interventionism and self-hygiene—*and* the middle-class perception of their lagging character—explains the choice that the British "cameralist" sanitarians embedded in the pre-1840s Improvement Acts. As E. P. Hennock noted, the Acts concerned themselves with the amenities and the public-space improvements that aimed for "comfort of the wealthier citizens." E. P. Hennock, "Urban Sanitary Reform a Generation before Chadwick," *Economic History Review* 10 (1957): 113–120, 117.

72. The Rev. Richard Price, "Observation on the Difference between the Duration of Human Life in Towns and Country Parishes and Villages," *Philosophical Transactions* 65 (1775): 424–445, 428.

73. *Hippocrates's treatise*, 4–5.

74. J. K. Walker, "Medical Topography of Huddersfield," *London Medical Repository* 10 (1818): 1–14.

75. Colin Chisholm, "On the Statistical Pathology of Bristol and Clifton," *Edinburgh Medical Journal* 13 (1817): 265–293.

76. James Johnson, *Change of Air or Pursuit of Health* (London: S. Highley, 1839), 10.

77. La Caze, quoted in Roselyne Rey, "Vitalism, Disease and Society," in Roy Porter (ed.), *Medicine in the Enlightenment* (Rodopi: Amsterdam, 1995), 465.

78. Dorothy Holland and Andrew Kipnis, "Metaphors for Embarrassment and Stories of Exposure: The not-so-Egocentric Self in American Culture," *Ethos* 22 (1994): 316–342, 330. See also Richard Shweder and Edmund Bourne, "Does the Concept of Person Vary Cross Culturally," in R. Shweder and R. LeVine (eds.), *Culture Theory* (Cambridge: Cambridge University Press, 1984).

79. William Coleman, "Health and Hygiene in the Encyclopaedia: A Medical Doctrine for the Bourgeoisie," *Journal of the History of Medicine* (1974): 399–421; see also Antoinette S. Emsch-deriaz, *Tissot: Physician of the Enlightenment* (New York: Peter Lang, 1992).

80. James Mackenzie Bloxam, *The Climate of the Island of Madeira: and the Errors and Misrepresentations of Some Recent Authors on the Subject* (London, T. Richards, 1854), 92.

CHAPTER 2

A Shift of View:

Meteorology in John Herschel's Terrestrial Physics

GREGORY A. GOOD

In his meteorological research, John Herschel (1792–1871) subjugated the intimate to the universal. Practitioners of a new approach had already usurped the dominance of the ancient, Aristotelian "meteoric tradition" which focused on unusual atmospheric events. A different meteorology based on physical theory and exhaustive, careful measurement prevailed. Theodore Feldman has examined this transition internationally, "from a literary, qualitative pursuit . . . to a quantitative subject based on abstract, mathematical theory and precise, systematic experiment"; James Fleming documented the emergence of this tradition and the interplay of theory, observation, and institution building in America; and Valdimir Jankovic characterized the practitioners of the new tradition as treating the entire atmosphere as a laboratory.[1] Herschel entered the scene just as this approach captured the field, conducting meteorological research during a transformative period for the science, from about 1820 to 1870. He shared the urban and laboratory backgrounds of John F. Daniell (1790–1845), Heinrich Dove (1803–1879), Joseph Henry (1797–1878), and others, just as he shared their commitment to the study of weather as a physical, global phenomenon.[2] Herschel simultaneously saw weather from a general, integrative perspective and supported analytical reduction of atmospheric phenomena to particular physical causes and mathematical laws.

This chapter begins by showing the place of Herschel's meteorology in his general understanding of natural philosophy. From early in his career, Hershel articulated a methodology that stressed causal explanation and the interconnections of physical phenomena, including those of meteorology and terrestrial physics. He explained atmospheric phenomena in terms of physical and chemical properties of matter. Herschel joined other influential contemporaries in explaining weather phenomena by such general properties.

Part II looks more closely at Herschel's position regarding quantification in meteorology, and specifically his work in actinometry. He held up measurement and mathematically stated laws as ideals for terrestrial physics generally, exemplifying how intimately small processes fit into his universal science.

Part III examines Herschel's methodological commitment to theory in meteorology through his effort to explain the movement of weather and winds. Specifically, Herschel collaborated with William Radcliffe Birt (1804–1881) to develop and test a theory of "atmospheric waves." Far from pursuing a Baconian or Humboldtean data-gathering exercise, Herschel emphasized the necessity of theory in meteorology just as he did in astronomy or physical optics. I contend here that Herschel's support of theorizing in meteorology paralleled how he believed all science operates.

Lastly, Part IV turns to John Herschel's advocacy of coordinated, global observation. He advocated not just a personal, intimate familiarity with all the particulars of global meteorological phenomena, but a communal intimacy. He argued that to bring the smallest matters of terrestrial physics within the grasp of universal science required scientists to collaborate, collectively becoming intimates of that encompassing knowledge. This required cooperative observation if the community was to develop synoptic (intimate, universal, and timely) views of phenomena. By articulating arguments for measuring meteorological variables, for boldly theorizing, and for broadly collaborating in research, Herschel promoted the institutionalization of meteorology as part of a wider terrestrial physics. Herschel provided means and justifications for reducing the local to a "specimen of the global." As Jankovic states it, "The culture of 'country airs' lost out to the physics of planetary circulation."[3] Herschel contributed to this new agenda, for meteorology and for terrestrial physics more generally.

TERRESTRIAL PHYSICS AND
NATURAL PHILOSOPHY

John Herschel practiced meteorology in a tradition of terrestrial physics, which saw geo-phenomena within the purview of natural philosophy.[4] Although some scientists had long treated some phenomena from this perspective—Newton the tides, Halley magnetic declination, Marsigli ocean waves—the possibility of a systematic development of terrestrial physics emerged in the early 19th century. Herschel devoted himself to promoting and embodying this science as much as he did to astronomy and optics. Indeed, Herschel came to terrestrial physics from natural philosophy. Through research articles and monographs, through popular and reflective essays, Herschel exercised significant sway over other scientists' views of research methods and topics. This section discusses how Herschel used these publications to position meteorology and terrestrial physics within natural philosophy.

In his *Preliminary Discourse on the Study of Natural Philosophy* (1830), Herschel explored using lessons learned in natural philosophy to study Earth and its environment.[5] He often illustrated difficult points about research methods with examples related to geology, the oceans, or the atmosphere. In this, his first published account of terrestrial physics or physical geography, Herschel called it "an extensive designation," which included climates, tides, magnetic variations, and geological structures.[6] It included mineralogy, the behavior of plants and animals, and "a thousand other particulars essential to that complete acquaintance with our globe as a whole. . . ."[7] Herschel championed physical study of the Earth. All sciences, he wrote, progressed from naming and classifying to seeking laws and causes, from natural history to natural philosophy.[8] The causes that operated on Earth and its fluid envelopes were the same as those studied in the laboratory. Laws governing these causes transcended laboratory and field.[9] Understanding meteorology meant understanding how "aërial fluids" behave in all pressures, temperatures, etc.[10] Pascal's experimental proof that air has weight led via the laboratory studies of Boyle, Hooke, and Mariotte to "a satisfactory knowledge of the general law of the equilibrium of the air under the influence of greater or less pressures."[11] Physical geography demonstrates the applicability of laws "on the great scale." Physical geography, including meteorology, reveals the "mutual relations and interagencies" that hold together the interacting parts of this world.[12]

Herschel selected atmospheric examples to illustrate two of the most important steps in scientific research in the *Preliminary Discourse*. He chose the formation of dew to show how an investigator conducts a research through the process of observation, induction, and the establishment of causes.[13] Herschel saw this as an instance of careful, experimental, meteorological research, in which an investigator related an unexplained phenomenon to two more general phenomena: the radiation of heat and the condensation of water vapor by cold. Dew forms when a surface cools by radiation "faster than its heat can be restored to it." The surface becomes colder than the air and water vapor condenses on it. Herschel, meanwhile, nimbly transitioned back and forth, linking lab results with field experience. He compared volcanoes with blowpipes, atmospheric winds with air currents in laboratory experiments, lightning with electric sparks, and earthquakes with vibrations in wires. He concluded that laws found in the laboratory are useful in studying the globe and cosmos.[14]

Herschel's second example connected terrestrial with cosmic phenomena: What qualifies as a true cause of the gradual, secular cooling of Earth's climate, which researchers thought to be occurring?[15] He excluded two candidate causes: that Earth was cooling from an original molten state and that more volcanoes in earlier times had warmed the climate. No one had shown that either of these conditions had ever actually prevailed. These ideas were purely speculative.

Herschel thought that two other causes qualified. While Herschel was writing the *Preliminary Discourse*, Charles Lyell (1797-1875) published the first volume of *Principles of Geology* (1830), including two chapters on past climates and climate change.[16] Lyell and Herschel both attended Geological Society meetings that year and bounced ideas off each other. Herschel noted in the *Preliminary Discourse* Lyell's argument that the change of continent and ocean placement affects climate. Since researchers had demonstrated that oceans erode continents and ocean bottom becomes dry land, Herschel accepted Lyell's mechanism as "a cause on which a philosopher may consent to reason. . . ."[17] Lyell inspired Herschel to pursue the second candidate: an astronomical cause. Lyell observed that precession of the equinoxes implied a slow exchange of relations to the Sun for the northern and southern hemispheres.[18] Herschel suggested that slow change in Earth's orbit caused Earth's cooling.[19] Secular increase of the minor axis of Earth's orbit meant less solar radiation reached Earth.[20] Hence, both Lyell's and Herschel's proposed causes existed and both could produce cooling. Further research would test whether they could produce the amount of cooling observed.

An important part of the *Preliminary Discourse* discusses the agency of heat in the atmosphere. Here Herschel related heat and gases. He noted that rapid condensation of gases produces heat and that rapid evaporation from a liquid into a gaseous state produces cold.[21] He approved the research of John Dalton (1766–1844) and J. L. Gay-Lussac (1778–1850) on the law relating temperature change to volume change. He noted that the law holds equally for all gases and observed that investigators had liquefied many solids and vaporized many liquids. Likewise, they had condensed most "aëriform fluids" into liquid by adequate cold and pressure.[22] In these phenomena, Herschel saw roles for heat in producing weather. Herschel saw the weather as balanced between the elasticity and compressibility of gases, between heat and gravity.[23]

Herschel outlined two main ways in which heat and gases interact. Heat acting similarly to light, radiates across great distances, and passes through gases. Earth's surface absorbs the Sun's heat and re-emits it. The atmosphere absorbs this re-emitted heat more readily than the Sun's direct heat. Edmé Mariotte (1620–1684) and C. W. Scheele (1742–1786), Herschel wrote, discovered this in laboratory experiments.[24] Pierre Prevost (1751–1839) established that all bodies simultaneously receive and radiate heat, which struck Herschel as especially important for meteorology. He used this idea to explain the formation of dew.

Secondly, Herschel said, Joseph Black (1728–1799) observed that when a liquid changes into a gas, the gas is no hotter than the liquid, meaning that much of the heat applied "has entered into the substance." Herschel in 1830 still considered this discovery of latent heat, along with the discovery that different materials have different specific heats, as evidence that heat can act similarly to a material substance.[25]

Eighteen years later, Herschel reviewed the first volume of *Kosmos* (1845) by Alexander von Humboldt (1769–1859) and its English translation (1846) by Elizabeth Leeves Sabine (1807–1889), wife of Edward Sabine (1788–1883).[26] This review provided Herschel an opportunity to comment on physical geography. After quoting Humboldt's distinction between geology and physical geography, Herschel concurred that the domain of physical geography was the "description of the actual state of the earth's surface in its three great divisions,—those of land, sea, and air . . . ."[27] Here Herschel used the word "actual" as in the French *actuel*, or "at the present time," current. In a sense, he referred to an eternal present, that is, he sought the ways that unchanging physical laws, applied to changing conditions, explain terrestrial phenomena of any present. The tacit distinction by Humboldt and the explicit distinction by Herschel between physical and historical investigations applied mainly to rock strata.[28] His discussions of oceanic and atmospheric phenomena emphasized physical explanation.[29]

Herschel published two book-length articles in the *Encyclopaedia Britannica*: "Physical Geography" (1859) and "Meteorology" (1857).[30] During the 1840s and 1850s, Herschel had been closely involved in terrestrial physics. Looking back, Herschel stated that the laws of physics underlay physical geography.[31] He demonstrated the applicability of these laws "on the great scale." Physical geography laid bare the "mutual relations and interagencies" that held together the particulars of geo-phenomena. Herschel thought physical geographers must know the laws of physics, at least well enough to trust what specialists in relevant areas concluded.[32]

Herschel illustrated the reliance of physical geography on physics with examples drawn from astronomy and dynamics.[33] From astronomy came a string of discoveries, including Earth's oblate, spheroidal figure, its volume, weight, and density. Herschel concluded from these results that Earth's innermost materials were either metallic or were condensed under extreme pressure. He concluded, based on temperatures in mines, volcanoes, and hot springs, and the fact that Earth's mean density is only 5.5, that Earth's interior must be hot enough to "counteract" the "condensing effect" of pressure. Two other conclusions about Earth's interior came from astronomy, particularly from lunar theory: Earth's center is under extreme pressure and density increases towards the center by "increasing proportion."

Herschel presented his best expression of a general view of meteorology in his book of that name. Here Herschel emphasized two sides of this science. Meteorology, first, was part of dynamical science.[34] If the atmosphere were not mobile, one could calculate the effects from the causes. Mobility, however, made atmospheric phenomena much more complex and led to "mechanical difficulties of a very high order." He lamented that phenomena of motion in elastic fluids were among the most resistant to dynamical treatment, even when the causes were "known and calculable." He meant the effects of heat on gases and the role of water vapor. The forces acting on the air, however, were not so quantifiable. Mitigating factors such as the irregular forms of continents and the presence of mountains made direct calculation impractical.

Herschel concluded that meteorology could not be a strictly deductive science. Meteorology had to rely on induction. Herschel wrote that this was most peculiar, since the most important efficient causes of the weather were well known. Only the "derivative causes" complicated the phenomena.

Herschel recommended that meteorologists, guided by both observations and dynamics, should focus their attention on "subordinate or derivative laws."[35]

In summary, John Herschel forcefully promulgated this new meteorology. He portrayed the atmosphere as a physico-chemical fluid, which followed mathematical laws. He emphasized the importance of measurement, careful observation, and mechanical theory. He reinforced abandonment of earlier attachment to place and immediate experience. He advocated a curious combination of an expansive universalism and a sense of the detailed interactions of Daltonian atomic events and force laws.

Through his discussion of meteorological phenomena as models of mechanical philosophy, Herschel simultaneously detailed the nuances of a new approach to meteorology and to terrestrial physics generally and lent his credibility to its methodology and explanatory standards.

MEASURING ATMOSPHERIC PHENOMENA

John Herschel's meteorology required new instruments and measurements of new physical variables. Meteorology, like terrestrial physics generally, addressed subtle phenomena. Whether observed like gravity with a swinging pendulum or like magnetism with a compass needle, this work required precision instruments and trained observers. The new meteorology required people to see an invisible mixture of substances, in which the laws of mechanics reigned, and in which heat and moisture played significant roles. Herschel helped direct atmospheric research toward careful and complete observation and toward quantitative measurements whenever possible. It seemed to Herschel that no one yet understood the solar radiation that drives the atmosphere, so he developed an instrument—the actinometer—to investigate radiation (Figure 1). This section sketches Herschel's general ideas about quantitative research and then focuses on this model instrument and Herschel's solar radiation program.[36]

When John Herschel toured the Continent as a young man in the 1820s, he traveled as a physical geographer. These scientific excursions incidentally prepared him to be a 'metropolitan' coordinator of scientist-expeditionaries and of far-flung, colonial, physical observatories. These excursions familiarized him with the research challenges, the instrumentation, and some of the hardships of the field. However, he traveled the Continent driven by his own ambitions and with the advantages of wealth and status. He visited scientists in Paris, Milan, and elsewhere, and he carried

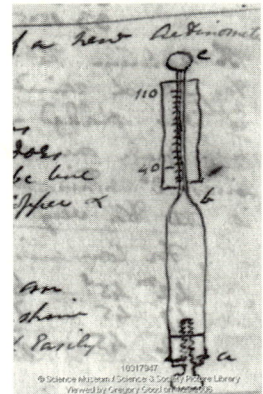

FIGURE 1   John Herschel's drawing of an
actinometer, c. 1830s, experiment 969 from Herschel's
notebook. Courtesy of Science Museum, London,
Science & Society Picture Library 10317947.

instruments. His three favorite instruments were the camera lucida, barometer, and actinometer.

On his first tour in 1821, made with Charles Babbage (1792–1871) and
two other friends, Herschel carried the following instruments:

A mountain Barometer
A small chest of Chemical apparatus
A Tennant's Blowpipe with the usual appendages
A Pocket Compass
A Small Pocket Sextant
A Good Thermometer
A large & small Mineralogical hammer
A reflecting Goniometer
A Camera Lucida.
Drawing board & other apparatus for drawing with various other use-
ful apparatus[37]

His interests were clear: altitudes, chemical analyses of minerals, exact positions, meteorological phenomena, mineralogy, and geological views. During
this tour and all his others, Herschel used the camera lucida to render careful sketches of many scenes, but especially of mountains.[38] The camera lucida symbolizes Herschel's use of other instruments to capture the essential
details, the likeness, of nature. The camera lucida afforded realistic views
of landscapes, clouds, etc., while the pictures provided a sense of awe. Herschel's other instruments similarly combined precision with scale, measured
number with theoretical concept.

Intimate awareness began for Herschel before making any measurements. He wrote in his *Preliminary Discourse* that a disciplined observation of a lightning strike required "exact notice" of appearance, sounds, smells, feelings of shocks, and tastes; also, presence of conductors, barometric pressure, temperature, and "the disposition of clouds."[39] Later, he recommended a "skeleton form" to aid in using the cloud classifications of Luke Howard (1772–1864), as a way to discipline and direct observers.[40]

Herschel explained in 1830 why measurement is critical to science.[41] To trust only to our impressions leads often to under- and over-estimation. Numerical precision "is the very soul of science" and allows us to test our theories. Since our senses are less trustworthy in estimating distance or time, "we are obliged to have recourse to instrumental aids."[42] Through the pendulum we measure time precisely; through Borda's repeating circle, distance. Herschel modeled and explained this essentially quantitative aspect of terrestrial physics in both his research and in his writing.

Sometime between 1821 and 1824—probably 1823, but the evidence is thin—Herschel developed his new instrument, the actinometer, to measure the intensity of solar radiation.[43] During the 1821 tour, Herschel and Babbage had climbed a high mountain in the Alps—they thought it was Monte Rosa, which they believed to be the highest mountain in Europe. They were wrong on both counts, but the mountain they climbed—the Breithorn—was high enough (4164 m, 13,661 feet) that Babbage especially suffered a terrible sunburn that laid him up for several days.

This inspired Herschel. An instrument that measured solar radiation would be useful in several inquiries auxiliary to the geology that brought Herschel to the mountains.[44] First, measurements of solar radiation on mountain peaks approximated the radiation incident on Earth's atmosphere from the Sun. Second, comparative measurements made at lower elevations revealed how much radiation is absorbed. These together approximated the radiation the atmosphere absorbs and explained meteorological phenomena. Herschel's actinometer measured the temperature increase in a deep-blue liquid per unit time caused by solar rays. This provided a relative measure of the intensity. The actinometer married Herschel's meteorological and physical interests.

Herschel depicted the use of the barometer and actinometer in several of his camera lucida drawings of his mountain travels. He produced hundreds if not thousands of such images, making Herschel the most accomplished and prolific camera lucida artist bar none. In a drawing made in the "Pass of the Jura at Champagnole Montagne Cornice" (probably in 1821), there

FIGURE 2   John Herschel's drawing of the Grotta dei Capri, Mount Etna, Sicily, showing the barometer he used to correlate altitudes with actinometer readings. His gaze was alternatively focussed on the barometer's mercury meniscus and the wild mountain scenery. The J. Paul Getty Museum, Los Angeles, John F.W. Herschel, Grotta dei Capri, Etna, 3 July 1824, graphite drawing made with the aid of a camera lucida, 19.8 cm × 31 cm, gift of the Graham and Susan Nash Collection.

is a small human figure (probably Charles Babbage) on the horizon using an instrument (probably a barometer).[45] Herschel employed the actinometer in 1824, while climbing Mount Etna and other mountains, to measure the absorption of solar radiation by the atmosphere at different elevations and showed it in use in No. 395 (not shown) and a barometer in use in No. 88 (Figure 2).[46]

Herschel shared his interest in solar radiation with several European scientists. During his 1824 tour he visited François Arago (1786–1853), Humboldt, and Joseph Fourier (1768–1830) in Paris. Fourier and Herschel particularly had arranged to meet.[47] Fourier had published his important article "Remarques générales sur les temperatures du globe terrestre et des espaces planétaire" in 1824.[48] These meetings allowed Herschel to ground his actinometric research in a more general European context.

Herschel did not publish an account of this instrument himself until ten years later.[49] The first published description of his actinometer was by William Ritchie (1790–1837), who had an interest in meteorology. Herschel and Ritchie exchanged detailed letters in 1825 on photometers, which prompted him to send Ritchie a description of his actinometer, since both instruments measure the intensity of radiation (one light, the other heat). In 1825, Ritchie quoted a long passage from one of Herschel's letters in an article on photometers.[50]

While Ritchie staked Herschel's claim to the instrument, Herschel quietly promoted its use. Through the 1820s and 1830s, the actinometer was a fixture in Herschel's correspondence. In 1826, he took a new version of the instrument to the top of the Puy-de-Dôme and other peaks, apparently trying to perfect it.[51] He began loaning copies of it to other scientists almost immediately,[52] especially to James David Forbes (1809–1868), who used it extensively in the 1830s during his studies of Alpine glaciers.[53] Forbes wrote to tell him in 1832 how impressed he was with Herschel's solar radiation work, which Herschel had described when Forbes visited Herschel's home in 1831.[54] Herschel replied:

> I shall take care (as the shortest & simplest thing I can do respecting the actinometers) to put a couple of them in working order and draw up a set of instructions for their use.[55]

Herschel suggested that if Forbes could set up "with some scientific friend" a set of observations at the base and summit of some mountain (he mentions Etna and Puy-de-Dôme) this would be "most valuable."

Herschel provided Forbes with a letter of introduction to Mario Gemmellaro (1773–1839) in Catania, Sicily, to promote the Mt. Etna trial of actinometers. Observations Herschel made in 1824 with Gemmellaro's assistance had failed because of a "new, imperfect instrumental aid," i.e., his first actinometer.[56] Herschel asked Gemmellaro this time to observe solar radiation at home when Forbes was on Etna.[57] Herschel explained his reason for the trial to Forbes, saying:

> The ascent of high mountains is of course (in the absence of astronomical trials, which would be the *best* mode of attacking the subject) the only way of accomplishing this object.[58]

Herschel continued with a detailed analysis of the instruments and listed test results to indicate their "degree of reliance." He assured Forbes that "the language they speak is definite & self-consistent." He concluded this important letter by explaining why an instrument for measuring terrestrial radiation differs from solar actinometers. Herschel saw the need to compare solar radiation, terrestrial radiation, and their absorption by the atmosphere. Understanding these phenomena would provide an essential background for meteorology.

In Herschel's 1840 report to the Royal Society on behalf of its Committee on Physics and Meteorology, he gave extensive directions for the use of the actinometer.[59] Herschel provided a similar description in the *Admiralty Manual* in 1851.[60] These matter-of-fact descriptions of instrument construction and use promoted uniform practices. He made the actinometer a standard instrument in the British, worldwide system of meteorological observation. Clearly, however, scientists from Continental Europe also investigated solar radiation and Earth's interior heat—Ludwig Kämtz (1801–1867) in Germany and Russia, Macedonio Melloni (1798–1854) in Italy, and Fourier and C.-S.-M. Pouillet (1790–1868) in France, for example. Although different researchers approached these phenomena through different instruments or questions, they shared a commitment to a cosmic view of Earth. All their research adhered to a quantitative, thermodynamic program. Although the French called the program *physique du globe*, the English and Italians terrestrial physics or *la fisica terrestre*, and the Germans *theoretische Erdkunde*, their practices agreed in broad terms.[61]

The actinometer represented the combination of close involvement with meteorological phenomena in lab and field with the goal of explanation based on universal physical principles. Without exact physical data, Herschel wrote, theories are "little better than mere inapplicable forms of words."[62] The new meteorology was to be a quantitative science.

THEORETICAL SPECULATIONS: GLOBAL CIRCULATION, HURRICANES, AND ATMOSPHERIC WAVES

John Herschel's commitment to mathematically expressed physical theory, such as the wave theory of Augustin Fresnel (1788–1827), is well known, but this commitment has not been set in the context of his physical geography. His insistence on instruments, measurement, and data sounds clearly empiricist, perhaps even naively inductive or Baconian. The presence of an engraving of Bacon and a quote from him on the title page of the *Preliminary Discourse* in 1830, indeed, reinforced this false impression. Herschel had tried to have these removed while the book was in press, but he was too late, thus strengthening his public image as an empiricist.[63] Although Herschel admired Bacon and realized the importance of the factual basis of meteorology and terrestrial physics, he insisted on the necessity to frame theories. This section traces Herschel's attention to meteorological theory from 1830 to the 1860s.

In the *Preliminary Discourse*, Herschel strongly endorsed hypotheses as a means to furthering generalization.[64] Hypotheses allow us to move far beyond the individual facts used to generate laws. Hypotheses propose explanations and analogies, which gradually develop structure and become theories. Such theories, he said, are "creatures of reason rather than of sense." He did not mean, though, that one could freely imagine theories:

> The liberty of speculation which we possess in the domains of theory is not like the wild licence of the slave broke loose from his fetters, but rather like that of the freeman, who has learned the lessons of self-restraint in the school of just subordination.[65]

That is, theorizing follows rules.

Herschel listed just two steps in building a theory: consider which agents or causes may help produce the phenomenon, and second, consider which laws might regulate the action of these agents. The philosopher must show that an agent acts as a *Vera causa* or true cause: it must exist and act according to natural laws.[66] A researcher must have "good inductive grounds" to believe the cause exists and that it behaves as posited. Herschel mentioned other criteria, too. One must derive laws independently "by direct induction" and subordinate assumptions must agree with experience. In the ultimate test, predictions calculated using a theory must conform to observations.[67]

Herschel illustrated these points with a series of theories that had succeeded: Coulomb's and Poisson's theories of electricity and magnetism, gravitation and planetary motion, and the wave theory of light. A theory must be bold, it should indicate clearly how to deduce results, and one must "try the truth of it" by comparing its consequences with facts.[68] If the tests succeed, one might have nature's true mechanism, a close approximation, or a good heuristic. Herschel concluded that we "cannot refuse to admit" a theory at least provisionally, since its success in explaining known phenomena indicates that it will likely explain others.[69]

Nevertheless, he was considering theories of the atmosphere. Recall his orbital-decay theory for secular cooling of the Earth (December 1830). Herschel made his understanding of the mechanics of the atmosphere more explicit in *A Treatise on Astronomy* in the Cabinet Cyclopædia; Herschel included a chapter on geography, where he discussed other atmospheric theories.[70] Here, in a general discussion of the relevance of geography to astronomy, he explained the trade winds. The uneven heating of Earth's surface causes the warmest, lightest air near the equator to rise, and colder air flows in low from the poles. Moreover, while Earth's rotation deflects air flowing toward the equator to the west, the warm air flowing toward the pole at higher elevations deflects toward the east.[71]

Herschel was well aware of Edmond Halley (1656–1742) and George Hadley's (1685–1758) explanations of the trade winds. Although he used Hadley's ideas, he based his discussion in *Treatise on Astronomy* on a more complete understanding of the behavior of gases than was available to Halley and Hadley. He increased his empirical knowledge of winds around the globe as well by assiduous reading of travel literature and through association with the Admiralty. He wrote:

> These mighty currents in our atmosphere . . . arise from, 1ˢᵗ, the unequal exposure of the earth's surface to the sun's rays, . . . and, 2dly, from that general law . . . of all fluids, in virtue of which they occupy a larger bulk, and become specifically lighter when hot than when cold.[72]

He combined this with Earth's rotation to explain a series of "observed facts." This was the test of theory. Herschel relied on "the general laws of hydrostatics" to trace out details, one by one, of the flow of air in the tropics and beyond: higher, warmer air currents traveling poleward and drifting east; lower, cooler air currents traveling from the poles toward the equator and drifting west. These "aërial currents," he said, produce the "two great tropical belts" of the trade winds.[73] In comparing the trade winds with the belts of Jupiter and Saturn, Herschel stressed that all of these arise from the combination of the planets' rotation and an external cause.[74]

Herschel thought scientists understood the basic principles of this circulation. He expressed himself somewhat differently in a letter to James

Hudson (1804–1859), then assistant secretary of the Royal Society, in 1835. He wrote:

> The physical cause . . . consists in the *upward suction*, which is the immediate consequence of the overflow of the equatorial atmospheric column into the extra-tropical regions, and which is not immediately compensated by the under-current of the Trades. It is a dynamical result, into which time enters as an essential element.[75]

Over the coming decades, he re-used the discussion from the *Treatise on Astronomy* almost verbatim, merely polishing the account.[76] When Herschel published *Outlines of Astronomy* in 1849, he added that the thermal behavior of the atmosphere was a *Vera causa*, as proven by the perpetually lower air pressure at the equator, the equatorial depression.[77]

His ideas about storms, however, evolved significantly. In 1833, he said that the global circulation discussed above usually does not produce incredible winds. Because the atmosphere is a mere skin on the planet and because these currents move slowly, they gradually take up the motion of the solid Earth over which they travel.[78] Nevertheless, he floated a "theory of hurricanes" in the *Treatise on Astronomy*. He proposed a question:

> It seems worth enquiry, whether hurricanes in tropical climates may not arise from portions of the upper currents prematurely diverted downwards before their relative velocity has been sufficiently reduced by friction on, and gradually mixing with, the lower strata; and so dashing upon the earth with that tremendous velocity which gives them their destructive character, and of which hardly any rational account has yet been given.[79]

This idea of large downbursts of upper winds, however, did not stand. We examine his later treatment of this question below.

Herschel asked as well in this passage whether two such air masses, transferred suddenly downward and crossing each other, might not produce "a tornado of any degree of intensity." This hypothesis is remarkable from two viewpoints. First, Herschel proposed it just about the time the American storm controversy was heating up. The theories proposed by James Espy (1785–1860), William Redfield (1789–1857), and Robert Hare (1781–1858) differed markedly from Herschel's (and each other).[80] Herschel's contribution as a storm theorist has yet to be fully examined. Second, this notion of "conflicting" air masses is a hypothesis that Herschel later developed and never relinquished. I am not aware that anyone else has noticed its 1833 provenance.

In the *Meteorology* and in later revisions of *Outlines of Astronomy*, Herschel returned to the question of hurricanes and tornados. He had closely followed the theories and researches of Redfield, Dove, Henry Piddington (1797–1858), and William Reid (1791–1858).[81] Herschel portrayed hurricanes by 1857 as "vorticose" whirlwinds, cyclones usually 200 to 300 miles in diameter and sometimes larger, which can drift "leisurely enough" at 2 to 30 miles per hour along a track. Among the six characteristics of hurricanes, Herschel included their direction of rotation (counter-clockwise in the northern hemisphere, clockwise in the southern) with a central calm and, further out, violent winds.[82] Herschel based his explanation of hurricanes on Alfred Swaine Taylor's (1806–1880) application of Dove's 'Law of Rotation' "to this specific class of aerial movements."[83] Air, heated over some larger area, begins to rise, and other air flows in along Earth's surface to replace it. Earth's rotation causes the in-flowing air to circle the rising column counter-clockwise (northern hemisphere) and the column itself gradually strengthens as a "vortex or spiral eddy."[84] Hence, Herschel concluded, hurricanes can occur only in middle latitudes, since, in higher latitudes, there is inadequate heat and, near the equator, the rotational effect is missing.

Herschel's most controversial meteorological theory concerned "atmospheric waves." As Jankovic states in his exploration of the development of this idea from 1843 to 1850, Herschel was perturbed at the "factology" that seemed to be taking root in meteorology and other sciences, as well as at the emphasis many placed on the usefulness of science. Jankovic writes that the publications and correspondence of Herschel and his collaborator William Radcliffe Birt (1804–1881) "reveal the complex of interests that in the early years of the British Association created an increased demand for meteorological research . . . ."[85] I add that Herschel had long advocated both extensive observation *and* theorizing. This was not new for him in the 1840s, or even in 1830, but sprang from his earliest experiences with his father, William Herschel (1738–1822). He wrote to William Whewell (1794–1866) in 1837:

What you say about the framing and trying and rejecting [of] hypotheses as leading threads to truth . . . is excellently well said. I remember it was a saying often in my Father's mouth "Hypotheses fingo" in reference to Newton's "Hypotheses <u>non</u> fingo" and certainly it is this facility of framing hypotheses if accompanied with an equal facility of abandoning them which is the happiest structure of mind for theoretical speculation.[86]

Susan Faye Cannon claimed in 1978 that Herschel's meteorology was a very good example of him "taking a light, sophisticated view of hypotheses."[87] As I will show below, Herschel's good intentions did not prevent his maintaining a hypothesis beyond its time. Specifically, he supported atmospheric waves long after others ignored them.

Herschel had long been interested in atmospheric pressure as an indicator of meteorological phenomena. Anticipating in 1832 a possible trip to South Africa, he arranged for Thomas Henderson (1798–1844) to compare the Royal Society barometer with that at the Cape Observatory.[88] Herschel had a small barometer network in mind: He had Henderson take his instruments to the Cape and return them to London in 1833 for more measurements, then took the same instruments back to the Cape himself in 1834.[89] He wrote James Prinsep (1799–1840), in Calcutta, that the barometer had traveled for the "express purpose" of calibration.[90] Herschel and his old friend Edward Ryan (1793–1875) made another set of comparisons, and Ryan went to India to tie Prinsep's barometer into this net.[91]

Atmospheric circulation was on Herschel's mind. He had written in 1833 that "great fluctuations, of the nature of waves" exist in the atmosphere.[92] He told Prinsep that his "speculative reasonings" began during the voyage to the Cape. He proposed that an "alternate, annual transfer" of part of the atmosphere occurred between hemispheres. This could arise, he said, in seasonal differential heating of the hemispheres, producing a seasonal differential of atmospheric pressure across the equator, the "epochs of maxima & minima" at Calcutta and the Cape being exactly opposite. In January, the pressure in Calcutta exceeded that at the Cape by 0.4 inch of mercury, "a propellant power . . . urging steadily & constantly the air towards the South, and vice versa at the opposite season." Herschel continued that this could be no local effect, but must "communicate motion to immense masses of air." He concluded that the effect was so "conspicuous" that he considered the cause "fully established." Constant pressure differences produce constant winds, he wrote, periodical pressure differences produce periodical winds.[93]

Herschel passed on this theory to numerous correspondents: Whewell, Forbes, George Biddell Airy (1801–1892), Francis Baily (1774–1834), Friedrich Wilhelm Bessel (1784–1846), and others. He frequently discussed organization of barometer comparisons and advice on reading instruments.[94] Herschel wrote Bessel that he suspected an "annual transfer of atmospheric pressure" between hemispheres.[95] Herschel was not only observing the southern sky at the Cape. He was theorizing about global air movements.

In the 1840s, as Jankovic relates, Herschel took a decidedly theoretical turn regarding atmospheric pressure. He and Birt, whom Herschel had employed to reduce barometric readings, worked together more intensively. Herschel was now immersed in meteorological and other terrestrial physics investigations, as he was chairing (and writing reports for) the Committee on Physics and Meteorology for the Royal Society of London. He was especially interested in graphical representation of barometric observations.[96]

Herschel at first spoke of "barometric curves" with Birt.[97] These became "atmospheric curves" in 1842.[98] Jankovic shows that in 1843 Herschel spoke of barometric waves and considered both the form of a wave and the molecular movement embodied by the wind.[99] He speculated that atmospheric waves noticed by Birt might recur. In December 1845, Birt affirmed that the "great atmospheric wave" had returned. He traced it year after year in the 1840s. Herschel rejoiced that this case of his theory, the germ of which he first developed in 1833, was gathering more empirical support and elaborated its application to explaining storms.

Jankovic traces Birt's growing disaffection with the theory in the late 1840s and 50s. Birt argued that the wave-like form of the barometric curve did not imply a real wave in the atmosphere and that it particularly did not imply a molecular movement that could produce wind.[100] By 1856, he still mentioned atmospheric waves, but said no one understood them well and that he preferred to reserve comment.[101]

Herschel, however, did not pull back from the idea. In 1857 and 1861, he still included a six-page section on atmospheric tides and waves in his article on meteorology in the *Encyclopedia Britannica*.[102] Here he said that like the ocean, the atmosphere has waves. Atmospheric waves, "rendered sensible" by the increase or diminution of barometric pressure, differed from ocean waves because the density of air decreases with elevation and because atmospheric waves propagate through the entire atmosphere, whereas ocean waves are only on the surface. Herschel claimed that atmospheric waves are strongest at ground level. He explained an experimental model using several immiscible fluids of different densities and colors that mimicked "aerial" waves. The lowest stratum was the most disturbed.[103]

He distinguished these waves from an atmospheric "heat" tide and a "gravitation" tide. In the former, the Sun's heat dilates the air and elevates its lines of equal density on one side of Earth, and the night contracts the air and lowers these lines on the other side. Hence, this cause produces a bulge that follows the Sun, circulating round Earth once per solar day. The gravitation tide, he said, is analogous to an ocean tide with two maxima per day, but with the air rising only about 6 feet above its average. This produced a negligible barometric change of 1/130th of an inch. In 1861, Herschel added a note that the heat tide produce no variation in pressure, since even with dilation, the atmospheric column contains the same mass as before. He applied this phenomenon to affirm an oscillation of air from day-lit hemisphere to the night-side, across Earth's terminator. Calculating that the heat tide raises the atmosphere 363 feet on one side and depresses it 363 feet on the other, this produces an angle of 3.5 seconds at the terminator. Air gliding from day to night at a mere one mile per hour, would transfer large portions of the atmosphere back and forth. Moreover, because the Sun oscillates from tropic to tropic over a year, this transfers air across the equator with an annual period and with a much greater barometric variation than the diurnal heat tide. Hence, Herschel finally explained more fully the theory he first expressed to Prinsep over 25 years earlier.[104]

Having explained this, Herschel separated atmospheric waves from aerial tides:

> The atmospheric waves here considered are those which originate, not from the general movement of the whole body of the atmosphere, but from internal displacements, the result of winds diverted from their course, or of great local disturbances of temperature, due to a concurrence of circumstances which may be termed casual, forasmuch as we cannot trace their laws.[105]

The wave movements—their ground speed and direction, and their height—he said were "rendered sensible" from the movement of the barometer. He cited a case of a wave that crossed Europe at 13.62 miles/hour. Maximum and minimum barometric pressures of the wave differed by 0.75 inches, indicating a vertical movement of lines of equal pressure of about 700 feet. The recurrent "great November waves," which Herschel credited to Birt, were much grander, their barometric pressure difference sometimes reaching 2 inches and the breadth of the wave reaching 6000 miles. Did Herschel still believe that these waves somehow produced storms? Absolutely. Urbain Le Verrier (1811–1877), he said, had demonstrated that the "great Crimean storm" of 14 November 1854 "was part and parcel of this phenomenon."[106]

Herschel continued to discuss these waves in publications in the 1860s. In 1864, he explained production of the November wave in detail. He claimed that several causes exaggerate and delay the southwest wind normally experienced by Europe. First, there is the transfer of air and water vapor from the southern hemisphere that begins after the Sun passes the autumnal equinox. Secondly, the prevailing southwesterlies in Europe begin in South America and the Sun reaches the zenith above South America just after autumnal equinox. This coincidence throws up "great torrents of vapour and intensely heated air," which, combined with the retreat of the trade-wind zone southward with changing season, produces "great temporary confusion and disturbance in the winds themselves." Herschel claimed that the first cause, operating especially in the Pacific Ocean, produces a stronger than normal southwesterly wind against the mountains on the west coast of North America, "to be thrown up along the whole length of that range into a broad swell, propagated onwards as a wave across America and the North Atlantic into Europe." This, added to the usual wind, produced the November wave.[107]

Herschel always indulged in speculation, contrary to his common image as a Baconian inductivist. He based his theories on mechanics and on the properties of matter and insisted that they must meet a series of criteria that he had worked out very early in his career.[108] Although later researchers modified some of his theories greatly, or rejected or forgot them, his approach reinforced a general trend in the mid-19th century toward basing meteorological theories on dynamics and thermal properties of gasses, not to mention on systematic observation.

ORGANIZING GLOBAL RESEARCH IN METEOROLOGY AND TERRESTRIAL PHYSICS

From an early age, John Herschel combined the solitary life of a researcher with that of an organizer of institutions. In the 1810s, he helped establish the Cambridge Analytical Society and the Royal Astronomical Society to promote continental mathematics and astronomy in England. He tried to reform the Royal Society in the 1820s. At the same time, he began organizing small-scale meteorological collaborations. He started in 1832 by helping Forbes with actinometric research, Henderson with barometric measurements (detailed above), and he gradually moved deeper into global coordination of investigation. By 1847, Herschel had so firmly established his leadership that the Admiralty chose him to edit its *Manual of Scientific Enquiry*.[109] This section examines Herschel's efforts to coordinate meteorological research and his vision of meteorology as part of a larger enterprise of terrestrial physics or physical geography. Herschel saw that sciences like meteorology, terrestrial magnetism, and geodesy required a communal, intimate awareness that greatly amplified the personal dedication needed by laboratory sciences. Synoptic views of global phenomena could only be achieved by many scientists working as one.

As early as 1830, Herschel began to see meteorology as ready for the transition from individual to collective activity. In his *Preliminary Discourse*, he wrote that as a science achieves general laws, some scientific investigators could specialize in observation and others in theory. Meteorology had achieved some general laws and was ready for a division of labor.

> Meteorology, one of the most complicated but important branches of science, is at the same time one in which any person who will attend to plain rules, and bestow the necessary degree of attention, may do effectual service.[110]

Hadn't geology in the previous generation benefited from the activity of individuals who simply set aside unfounded theory for careful collection and observation? Indeed, Herschel proposed the development of "printed skeleton forms" (i.e., structured, printed documents with blanks for observed data) in the *Preliminary Discourse*. In this book, he introduced the call for a centrally controlled network that became hallmarks of the Magnetic Crusade of the 1840s and the *Admiralty Manual* activities in the 1850s, both of which Herschel helped coordinate.[111]

Herschel began organizing meteorological research in earnest just before he left England for the Cape of Good Hope, South Africa in 1833. Shortly after he arrived in Cape Town in January 1834, he helped Thomas Maclear (1794–1879) standardize the barometer and thermometer at the Royal Observatory.[112] By April, he was asking if the Cape colony's meteorological journal was still active.[113] In September, he wrote Whewell, describing his organizational efforts.[114] He was coordinating the Meteorological Committee of the South African Literary and Philosophical Society, spurring others to action, and writing the committee's reports. Herschel lamented that his colleagues were not "thinkers" but proposed that they could be made into observers. He requested that Whewell supply them with "stimulants" in the form of questions, as well as autocratic orders for measurements.

Herschel's program began to emerge. It included the obvious elements of phenomena (solar radiation, atmospheric absorption, barometric pressure, etc.) and instruments. It included Herschel's first efforts to manage a colonial network of observers.

To realize his scheme, Herschel wrote and printed a 17-page *Instructions for Making and Registering Meteorological Observations*.[115] This publication signaled Herschel's intent to formalize collaborative research. In these *Instructions,* he advocated "a concerted plan of contemporaneous observations." The scientific investigation of the laws of climate and meteorology required this data, "taken in conjunction with the known laws of physics."[116] He specified two types of observation schedules: a daily schedule of either three or (preferably) four observation times; and four days each year—the 21st of March, June, September, and December—with hourly observations. He explained the use of barometers, thermometers, maximum-minimum thermometers, and hygrometers. He suggested measuring soil and seawater temperature. He recommended how to observe the wind. He even included directions regarding atmospheric electricity and the tides, although the latter was not "strictly speaking, a branch of meteorology."[117] This betrays Herschel's broader goal of encouraging terrestrial physics generally.

Herschel sent a copy of these instructions to James Hudson at the Royal Society.[118] He confessed to Hudson that:

A sea voyage, and a residence in a climate every way so remarkable as this, has a powerful tendency to make a man a meteorologist, (at least *in wish*,) and I have, accordingly, been led to pay much more attention to the phenomena of the weather, and the indications of meteorological instruments, since leaving England, than ever I did before.

He described the four solsticial/equinoctial days of intensive, hourly measurements as "great meteorological festivals" when observers around the globe should be at their stations. Herschel was not the only one to suggest such coordination; Gauss, Humboldt, and Arago numbered among advocates of global research. However, Herschel's voice added extra luster to this cause.

As advocated in the *Instructions,* Herschel instigated collaborative meteorological work in other colonies. Throughout 1834 and 1835, he wrote numerous correspondents about barometric and thermometric investigations. He asked John Augustus Lloyd (1800–1854), in Mauritius, to make hourly observations.[119] He encouraged James Prinsep and Edward Ryan in Calcutta to investigate barometric pressure and atmospheric circulation in India.[120] Prinsep offered to collect data from across the sub-continent that might help Herschel's investigation. Many reports of data from observers were for March, June, September, and December, indicating that they were participating in Herschel's "meteorological festivals." He reported his own hourly equinoctial and solsticial meteorological observations in outlets in Cape Town and Britain.[121] Meanwhile, Herschel stayed in touch with Airy, Forbes, and others in Britain who were actively organizing meteorological research.

Although his time in Africa was heavily committed to astronomy, terrestrial physics (and meteorology particularly) were becoming ever more important to him. He was clearly forming his ideas about the importance of observatories dedicated to meteorology, tides, geomagnetism, and other earth sciences. He praised John Augustus Lloyd's "geographical observatory" as a crucial adjunct to astronomical observatories. He wrote Francis Beaufort (1774–1857), hydrographer to the Admiralty, defining what he meant by "physical observatories."[122]

When he returned to England in 1838, Herschel quickly became involved in the efforts to institute physical observatories around the globe. He played a critical role in convincing the British government to support the Magnetic Crusade.[123] This "crusade," however, transcended magnetism. Atmospheric instructions for the expedition of James Clark Ross (1800–1862) and colonial stations in Toronto and Hobart were broad. They included cloud classification, actinometry, barometric and thermometric measurements, and the use of radiating thermometers, hygrometers, anemometers, rain gauges, and electrometers.[124] As he said in his presidential speech to the British Association in 1845, terrestrial physics required support of organized observation and calculation just as much as astronomy did.

> Speaking in a utilitarian point of view, the globe which we inhabit is quite as important a subject of scientific inquiry as the stars. . . . Terrestrial Physics, therefore, form a subject every way worthy to be associated with Astronomy as a matter of universal interest and public support, and one which cannot be adequately studied except in the way in which Astronomy itself has been—by permanent establishments keeping up an unbroken series of observations . . . the gigantic problems of meteorology, magnetism, and oceanic movements can only be resolved by a far more extensive geographical

distribution of observing stations, and by a steady, persevering, systematic attack, to which every civilized nation . . . ought to feel bound to contribute its contingent.[125]

Herschel's role in establishing physical observatories and systematic terrestrial physics began in the 1830s with meteorology and continued in the 1840s and 50s with terrestrial magnetism. By the 1860s, physical observatories were solidly established, not only in Britain and its colonies, but across Europe, Asia, and the Americas. Collective investigation of the Earth took off.

CONCLUSION: JOHN HERSCHEL'S INTIMATELY UNIVERSAL TERRESTRIAL PHYSICS

In sum, John Herschel was involved in meteorology at every conceivable level, from most basic observation, through the invention and perfection of new instrumentation, to the development of general laws and theories and the encouragement of the institutional infrastructure required for such a data-intensive science. Herschel wed the urge of the Romantic to experience natural phenomena personally with the conviction of a cognoscente of celestial dynamics that phenomenon can ultimately be calculated from simple, universal laws. He climbed mountains to measure solar radiation unimpeded by atmospheric absorption. He followed long chains of analysis—both mathematical and not—to find the causes of dew and of gradual changes in Earth's climate over geological time. He looked forward to the day when researchers would integrate meteorological phenomena into dynamics and thermodynamics. He sought to unite terrestrial phenomena "in the wild" with discoveries made in the laboratory and in theory. And lastly, he sought to build a community of observers who would collectively produce a universal knowledge of the intimate details of Earth's many, varied phenomena. At the dinner that welcomed Herschel home from Africa in 1838, Roderick Murchison (1792–1871) proclaimed that for Herschel,

. . . not only the great phenomena of the universe, but . . . the minutest of its wondrous details were alike open to his mind . . .[126]

Herschel reflected many of the most important changes that were taking place in meteorology in the mid-nineteenth century, and in fact, he was an important actor in those changes. Herschel prominently promoted a vision

58

of terrestrial physics that brought the most intimately detailed observations within the dominion of the most universally expressed physical theory; the means to this subjugation was the coordinated, global cooperation of many individuals.

NOTES

1. Theodore Sherman Feldman, "The History of Meteorology, 1750–1800: A study in the quantification of experimental physics," Ph.D. dissertation, University of California, Berkeley, 1983. 286 p.; James Rodger Fleming, *Meteorology in America, 1800–1870* (Baltimore: Johns Hopkins University Press, 1990); Vladimir Jankovic, *Reading the Skies: A Cultural History of English Weather, 1650–1820* (Chicago: University of Chicago Press, 2000), ch. 7, "Laboratory Atmospheres," pp. 143–164. Jankovic argues that this tradition began in the 17th century and co-existed alongside meteoric 'reportage' until the early 19th century.

   I have presented elements of this article several times, but most notably at the conference "Science and British Culture in the 1830s," sponsored by the Royal Society of London at Cambridge, England, in 1994. Most of the argument, however, is new with this presentation. Final revisions have been supported by the National Science Foundation under grant 0432202.
2. John Frederic Daniell, *Meteorological Essays and Observations* (London: Thomas and George Underwood, 1823) and *Elements of Meteorology*, 3d ed., 2 vols. (London: Parker, 1845); and Heinrich Wilhelm Dove, *Meteorologische Untersuchungen* (Berlin: Sandersche Buchhandlung, 1837) and *Das Gesetz der Stürme in seiner Beziehung zu den allgemeinen Bewegungen der Atmosphäre*, 2d ed. (Berlin: Verlag von Dietrich Reimer, 1861).
3. Jankovic, *Reading the Skies*, 2000, p. 11.
4. See Gregory A. Good, "The Assembly of Geophysics: Scientific Disciplines as Frameworks of Consensus," *Studies in the History and Philosophy of Modern Physics* 31 (2000): 259–292.
5. John Losee provides a short overview of Herschel's ideas about induction, framing hypotheses, and forming theories in his *A Historical Introduction to the Philosophy of Science*, 4th ed. (Oxford and New York: Oxford University Press, 2001), pp. 104–108.
6. Herschel alternated between "terrestrial physics" and "physical geography" for the physical and chemical investigation of geo-phenomena. He contrasted this science with geology, which concerned itself with Earth's history. Physical geography concerned Earth's present.
7. John Herschel, *Preliminary Discourse on the Study of Natural Philosophy* (London: Longman, Rees, Orme, Brown, and Green, 1830), p. 350.

8. Herschel, *Preliminary Discourse*, 1830, pp. 75–220.

9. Herschel, *Preliminary Discourse*, 1830, p. 221.

10. Herschel, *Preliminary Discourse*, 1830, p. 228 f.

11. Herschel, *Preliminary Discourse*, 1830, pp. 230–231.

12. John F. W. Herschel, *Physical Geography* (Edinburgh: Adam and Charles Black, 1861). First published in *Encyclopaedia Britannica*, 8th ed. (1859), 17: 569–647.

13. Herschel, *Preliminary Discourse*, 1830, pp. 159–164. William Charles Wells (1757–1817) did this work.

14. Herschel, *Preliminary Discourse*, 1830, p. 173.

15. Herschel, *Preliminary Discourse*, 1830, pp. 145–148.

16. Charles Lyell, *Principles of Geology, being an Attempt to Explain the Former Changes of the Earth's Surface, by Reference to Causes now in Operation*, 3 vols. (London: John Murray, 1830-1833), especially chapters six and seven, 1: 92–124.

17. Herschel, *Preliminary Discourse*, 1830, p. 147.

18. Lyell, *Principles of Geology*, 1830, 1:110.

19. Herschel, *Preliminary Discourse*, 1830, pp. 147–148. John Herschel, "On the Astronomical Causes which may influence Geological Phenomena," *Transactions of the Geological Society (London)* 3 (1835): 293–299; summaries in *Geological Society Proceedings* 1 (1830–1831): 244–245 and *Philosophical Magazine* 9 (1831): 136–138. Herschel gives the date of this presentation as 15 December 1830 in his *Outlines of Astronomy*, American ed., 2 vols. (Reprinted New York: P. F. Collier & Son, 1902), 1: 17. He added the note in 1865. This American edition was based on the 1869 tenth edition. Herschel submitted his last revisions of the *Preliminary Discourse* to the press two days after his Geological Society presentation (Herschel manuscripts, Diary, 17 December 1830, TxU:H/W-0012. University of Texas, Humanities Research Center). I have adopted the archival abbreviations used in Michael J. Crowe, et al., eds., *A Calendar of the Correspondence of Sir John Herschel* (Cambridge, UK: Cambridge University Press, 1998).

20. James Croll (1821–1890) and Milutin Milankovitch (1879–1959) are better known than Herschel for their later investigation of the relations of climate and orbital elements. See James Rodger Fleming, *Historical Perspectives on Climate Change* (New York and Oxford: Oxford University Press, 1998). Fleming discusses these later events in more detail in: "James Croll in Context: The encounter between climate dynamics and geology in the second half of the nineteenth century," *Proceedings of the Milutin Milankovitch Anniversary Symposium: Paleoclimate and the Earth's Climate System*, André Berger, ed. (Belgrade: Serbian Academy of Arts and Sciences, 2005), pp. 11–18. Croll's first publication on climate and orbital parameters was: "On the Physical Cause of the Change of Climate during Geological Epochs" *Philosophical Magazine* 28 (1864): 121–137.

21. Herschel, *Preliminary Discourse*, 1830, pp. 313 and 318–319.

22. Herschel, *Preliminary Discourse*, 1830, pp. 320–321.

23. Herschel, *Preliminary Discourse*, 1830, p. 322.

24. Herschel, *Preliminary Discourse*, 1830, pp. 314–315. Mariotte and Scheele, he wrote, discovered this, in the laboratory.

25. Herschel, *Preliminary Discourse*, 1830, pp. 322–323.

26. Herschel, "[Review of] Humboldt's . . . *Kosmos*," *Edinburgh Review* 87 (January 1848): 170–229; reprinted in Herschel, *Essays from the Edinburgh and Quarterly Reviews, with Addresses and other Pieces* (London: Longman, 1857), pp. 257–364; and in *The Edinburgh Review or Critical Journal, for January and April 1848*, American ed., 87 (1848): 90–121. Citations are to the latter printing.

27. Herschel, "[Review of] Humboldt's *Kosmos*," 1848, p. 106.

28. Herschel, "[Review of] Humboldt's *Kosmos*," 1848, pp. 111–112.

29. Herschel, "[Review of] Humboldt's *Kosmos*," 1848, pp. 114–119.

30. Herschel, *Physical Geography* (1861); John F. W. Herschel, *Meteorology* (Edinburgh: Adam and Charles Black, 1861). First published in *Encyclopaedia Britannica*, 8th ed. (1857), 14: 636–690. *Physical Geography* had five editions and *Meteorology* had two. Unless otherwise remarked, Herschel wrote the passages quoted from *Meteorology* in 1857.

31. Herschel, *Physical Geography*, 1861, pp. 2–4.

32. Herschel, *Physical Geography*, 1861, pp. 3–4.

33. Herschel, *Physical Geography*, 1861, pp. 4–9.

34. Herschel, *Meteorology*, 1861, pp. 1–5.

35. Herschel, *Meteorology*, 1861, pp. 5–6.

36. See Adelheid Voskuhl, "Recreating Herschel's actinometry: An essay in the historiography of experimental practice," *British Journal for the History of Science* 30 (1997): 337–355.

37. John Herschel, Journal, 1821, TxU:H/W-0003. Herschel toured parts of Europe in 1821, 1822, 1824, 1826, and 1829.

38. Larry J. Schaaf, *Tracings of Light: Sir John Herschel & the Camera Lucida, Drawings from the Graham Nash Collection*, with introductions by Graham Nash and Graham Howe (San Francisco: The Friends of Photography, 1990). Herschel was the most productive camera lucida artist; hundreds of his drawings still exist.

39. Herschel, *Preliminary Discourse*, 1830, pp. 120–121.

40. Luke Howard, "On Clouds, Askesian Lectures, 1802," *Philosophical Magazine* 16 (1803): 97–107, 344–357, and 17 (1803): 5–11; and [John Herschel], *Report of the Committee of Physics, including Meteorology, on the Objects of Scientific Inquiry in those Sciences, approved by the President and Council* (London: Richard and John E. Taylor, 1840), 121 p., on pp. 73–74.

41. Herschel, *Preliminary Discourse*, 1830, pp. 122–134.

42. Herschel, *Preliminary Discourse*, 1830, p. 125.

43. The first mention is in a letter from Francis Lunn to John Herschel, 23 September [1823?], RS:HS 11.409. Although he specifies in 1859 that he first measured solar radiation during his tour of Sicily in 1824, he likely started work on the actinometer before then. See John Herschel, *Meteorology*, 1861, p. 12. Adelheid Voskuhl has recreated much of Herschel's own work in actinometry and through her own careful use of the instrument has demonstrated the challenges scientists faced in using it. She explores the historiographic issues of recreating an instrumental practice and what this activity can tell historians that documents cannot. See: Voskuhl, "Recreating Herschel's actinometry.

44. H. B. Saussure (1740–1799) was among the first to measure solar radiation in the Alps, in the 1780s and 90s. See: René Sigrist, ed., *H.-B. de Saussure (1740–1799): un regard sur la terre* (Paris: Georg, 2001); René Sigrist, *Le capteur solaire de Horace-Bénédict de Saussure: Genèse d'une science empirique* (Genève: Editions Passé-Présent, Librairie Jullien Éditeur, 1993); and Margarida Archinard, "The scientific instruments of Horace-Bénédict de Saussure," *Studies in the history of scientific instruments*, Christine Blondel, et al., eds. (London: Rodgers Turner, 1989), pp. 83–95.

45. Larry J. Schaaf, *Tracings of Light*, 1990. No. 339 is reproduced on p. 18. Of the 310 in the Nash collection, around 40 are reproduced in the book. Schaaf says the instrument in this view is an actinometer, but it looks more like a barometer.

46. Schaaf, *Tracings of Light*, 1990, p. 115. No. 395 is unfortunately not reproduced in the book, so one cannot be sure it includes an actinometer.

47. Fourier to Herschel, 6 April 1824, Royal Astronomical Society, William Herschel Collection 1/13.F.7.

48. Jean Baptiste Joseph Fourier, "Remarques générales sur les temperatures du globe terrestre et des espaces planétaire," *Ann. chem. phys.* 2nd ser., 27 (1824): 136–167. See Fleming, *Historical Perspectives on Climate Change*, 1998, chapter 5, "Joseph Fourier's Theory of Terrestrial Temperatures," pp. 55–64.

49. He first mentioned it in print in *A Treatise on Astronomy* (London: Longman, Rees, Orme, Brown, Green and Longman, and John Taylor, 1833), p. 210, note. "By direct measurement with the *actinometer*, an instrument I have long employed in such enquiries, and whose indications are liable to none of those sources of fallacy which beset the usual modes of estimation, I find that out of 1000 calorific solar rays, 816 penetrate a sheet of plate glass 0.12 inch thick; and that of 1000 rays which have passed through such plate, 859 are capable of passing through another." John Herschel, "Explanation of the Principle and Construction of the Actinometer," *British Association Report for 1833* (1834), pp. 379–381.

50. William Ritchie, "Additional Observations on Leslie's Photometer, etc." *The Edinburgh Journal of Science* 3 (1825): 104–107. The extract from Herschel's letter is quoted on p. 107. The Ritchie-Herschel correspondence spans 1825 to

1832. Most of the letters are at the Royal Society of London, with a few at the University of Texas.

51. See, for example, letter of John Herschel to William Henry Fox Talbot, 17 September 1826, Science Museum, Talbot 1/3.
52. Letter from Joseph Nicolas Nicollet, 9 August 1827, RS:HS 13.136.
53. Letter of John Herschel to James David Forbes, 23 May 1832, St. Andrews 73. Forbes and Herschel exchanged scores of letters between 1831 and 1859.
54. Forbes to Herschel, 10 May 1832, RS:HS 7.276.
55. Herschel to Forbes, 23 May 1832, RS:HS 21.110.
56. Herschel to Mario Gemmellaro, c. 28 June 1832 [?], St. Andrews 113.
57. Herschel to Gemmellaro, c. 28 June 1832 [?]. His wording, however, was ambiguous: "As corresponding obsns. are desireable (s:c) I hope he will have the advantage of your cooperation."
58. Herschel to Forbes, 23 May 1832, RS:HS 21.110. Forbes ended up meeting German meteorologist Ludwig Friedrich Kämtz (1801–1867). Together they made actinometric observations and Forbes loaned Herschel's instruments to Kämtz until they met again in 1833. See Forbes to Herschel, 28 April 1833, RS:HS 7.284. Kämtz was then writing his *Lehrbuch der Meteorologie*, 3 vols. (Halle: In der Gebauerschen Buchhandlung, 1831–1836).
59. John Herschel, *Report of the Committee of Physics, including Meteorology, on the Objects of Scientific Inquiry in those Sciences* (London: Richard and John E. Taylor, 1840), pp. 61–68.
60. John Herschel, ed., *A Manual of Scientific Enquiry; Prepared for the Use of Officers in Her Majesty's Navy; and Travellers in General*, 2d ed. (London: John Murray, 1851), pp. 299–309; hereafter, *Admiralty Manual*.
61. Good, "The Assembly of Geophysics," 2000, passim.
62. Herschel, *Preliminary Discourse*, 1830, p. 212.
63. Herschel to William Henry Fitton, 27 December 1830. This manuscript letter is privately owned.
64. Herschel, *Preliminary Discourse*, 1830, pp. 190–209.
65. Herschel, *Preliminary Discourse*, 1830, pp. 190–191.
66. See Gregory Good, "John Herschel's Optical Researches and the Development of his Ideas on Method and Causality," *Studies in the History and Philosophy of Science* 18 (1987): 1–41. The analysis of *verae causae* runs from pp. 21 to 26.
67. Herschel, *Preliminary Discourse*, 1830, pp. 197–198.
68. Herschel, *Preliminary Discourse*, 1830, p. 199.
69. Herschel, *Preliminary Discourse*, 1830, p. 209.
70. John Herschel, *Treatise on Astronomy*, 1833. Chapter 3 "Of Geography" discusses determination of position, trigonometric surveys, geodesy, cartographic projections, and measurement of elevation with a barometer. He completed this book in May 1833: John Herschel to Margaret Brodie Herschel (spouse), 22 May 1833, JHS.HCEJ.523. Private collection. Although Herschel claimed to

address only the relevance of geography to astronomy, he could not avoid discussing physical causes.

71. Herschel, *Treatise on Astronomy*, 1833, pp. 128–132. Herschel did not mention
G. G. Coriolis (1792–1843), whose work was published first in 1836. Whereas
Herschel and Hadley addressed only particular cases of the effect of Earth's ro-
tation on fluids, Coriolis developed a more general treatment.

72. Herschel, *Treatise on Astronomy*, 1833, p. 128.

73. Herschel, *Treatise on Astronomy*, 1833, pp. 129–132.

74. Herschel, *Treatise on Astronomy*, 1833, p. 212, note. This highly speculative
passage ties these phenomena with sunspots, aurorae borealis, Fraunhofer's
lines, and the absorption of heat and light by atmospheres.

75. The letter (dated 8 January 1835) does not survive, but Hudson printed extracts
of it as: [John Herschel], "Barometrical and other Observations by Sir John Herschel" *Athenaeum* no. 391 (25 April 1835): 320.

76. Herschel, *Outlines of Astronomy* (London: Longman, Brown, Green, and Longmans, 1849), paragraphs 239–245. Because Herschel preserved his numbered
paragraphs throughout the various editions of this book, I cite the paragraph
here.

77. Herschel, *Outlines of Astronomy*, 1849, paragraph 240.

78. Herschel, *Treatise on Astronomy*, 1833, p. 130.

79. Herschel, *Treatise on Astronomy*, 1833, note, p. 132. See also, William Whewell
to Forbes, June 10, 1833, in Isaac Todhunter, ed., *William Whewell, D. D., master of Trinity college, Cambridge. An account of his writings with selections
from his literary and scientific correspondence* (London: Macmillan and Co.,
1876), 2: 165–167. This letter is quoted in Katharine Anderson, *Predicting the
Weather: Victorians and the Science of Meteorology* (Chicago: University of
Chicago Press, 2005), p. 86: "Can any facts be collected to verify or test Herschel's theory of hurricanes? Again, will any analyst solve for us the problem of
the pressure of an elastic fluid *in motion*, so as to apply to the barometer?"

80. See Nathan Reingold, ed., *Science in Nineteenth-Century America: A Documentary History* (New York: Hill and Wang, 1964; Chicago: University of
Chicago Press, 1985), pp. 92–107; and James Rodger Fleming, *Meteorology in
America, 1800–1870* (Baltimore: The Johns Hopkins University Press, 1990),
chapter 2, pp. 22–54. Reingold provides a short introduction and reprints several important original letters; Fleming documents the entire controversy.

81. Henry Piddington, *Notes on the law of storms: as applying to the tempests of
the Indian and Chinese seas drawn up for the use of the expedition to China*
(Calcutta: G.H. Huttmann, Bengal Mily. Orphan Press, 1840) and *The Sailor's
Horn-book for the Law of Storms*, 2d ed. (London: Smith, Elder and Co., 1851);
and William Reid, *An attempt to develop the law of storms by means of facts,
arranged according to place and time; and hence to point out a cause for the
variable winds, with the view to practical use in navigation* (London: Weale,

1838). George Eden (1784–1849), Lord Auckland and First Lord of the Admiralty, sent Herschel extracts of works by Piddington and Reid when Herschel was working on the *Admiralty Manual*: Eden to Herschel, 11 Feb. 1848, RS:HS 7.13.

82. Herschel, *Meteorology*, 1861, pp. 66–69.

83. Herschel, *Meteorology*, 1861, p. 67; and Alfred Swaine Taylor, *On the temperature of the earth and sea, in reference to the theory of central heat: a lecture* (London, 1846).

84. Herschel, *Meteorology*, 1861, p. 69. In the 1860s, Herschel added notes on hurricanes to his *Outlines of Astronomy* expressing the same theory, paragraphs 245a to 245c.

85. Jankovic, "Ideological Crests versus Empirical Troughs," *British Journal for the History of Science* 31 (1998): 21–40, on pp. 21–22.

86. Herschel to Whewell, 20 August 1837, RS:HS 21.228.

87. Susan Faye Cannon, "Humboldtian Science" in *Science in Culture: The Early Victorian Period* (New York: Neale Watson Academic Publications, 1978), pp. 93–95.

88. Thomas Henderson to Herschel, 13 January 1832, RS:HS 9.284. Francis Beaufort arranged for Henderson to meet Herschel to discuss cooperative investigations. Beaufort to Herschel, 2 January 1832, RS:HS 3.325.

89. Henderson to Herschel, 27 April 1833, RS:HS 9.287. Herschel arrived at the Cape in January 1834.

90. Herschel to James Prinsep, 2 December 1834, RS:HS 25.54.4. Prinsep was assay master at the Indian mint.

91. Herschel to Edward Ryan, 28 October 1834, TxU:H/L-0765.1. Ryan was a judge in India.

92. Herschel, *Treatise on Astronomy*, 1833, pp. 25–26.

93. Herschel to Prinsep, 2 December 1834, RS:HS 25.54.4. Herschel discussed the "equatorial depression" of the barometer in this letter and called it "a general & not a casual phænomenon." Herschel acknowledged in his *Meteorology* that Humboldt and Joachim Frederik Schouw (1789–1852) were the first to observe the equatorial depression (p. 54).

94. Herschel to Whewell, 21 September 1834, RS:HS 25.3.21; Forbes to Herschel, 5 February 1835, St. Andrews, LB II, 215–217; Herschel to George Biddell Airy, 19 October 1834, RS:HS 1.66; and Herschel to Francis Baily, 22 October 1835, RS:HS 25.4.16 and 25.8.7.

95. Herschel to Friedrich Wilhelm Bessel, 16 May 1835, RS:HS 25.4.10.

96. Herschel to Birt, 22 May 1840, RS:HS 19.79 and 80.

97. Herschel to Birt, 16 March 1841, RS:HS 19.88.

98. Herschel to [Birt], 8 June 1842, RS:HS 19.95.

99. Herschel to Birt, 28 July 1843, RS:HS 19.106, quoted in Jankovic, "Ideological Crests versus Empirical Troughs," 1998, p. 29, note 34.

100. Jankovic, "Ideological Crests versus Empirical Troughs," 1998, pp. 36–37.

101. Jankovic, "Ideological Crests versus Empirical Troughs," 1998, p. 38. William Radcliffe Birt, *Handbook of the Law of Storms; being a Digest of the Principal Facts of Revolving Storms for the Use of Commanders in Her Majesty's Navy and Mercantile Marine* (London, 1856), p. iv.

102. Herschel, *Meteorology*, 1861, pp. 69–76.

103. Herschel, *Meteorology*, 1861, pp. 69–70.

104. Herschel, *Meteorology*, 1861, pp. 71–73.

105. Herschel, *Meteorology*, 1861, p. 74.

106. Herschel, *Meteorology*, 1861, pp. 74–76.

107. Herschel, "The Weather, and Weather Prophets," *Good Words*, January 1864, 5: 57–64, reprinted in Herschel. *Familiar Lectures on Scientific Subjects* (London: Strahan, 1866), pp. 142–175. Citations are to the latter printing. Atmospheric waves are discussed on pp. 163–167. His other publications on these waves are: "[Letter on Barometric Waves]," *Manchester Literary and Philosophical Society Proceedings* 6 (1867): 91–93; and "Le vent et le baromètre," *Revue Maritime et coloniale* 19 (1867): 927–928.

108. See, for example, Good, "Herschel's Optical Researches," 18 (1987): 1–41.

109. See George Eden to Herschel, 19 November 1847, RS:HS 7.9; and Herschel to Eden, 20 November [1847], RS:HS 23.51. Herschel *Admiralty Manual*, 1851. Herschel wrote the original Preface (pp. v–ix) to the *Manual* and the article "Meteorology" (pp. 280–336).

110. Herschel, *Preliminary Discourse*, 1830, p. 133.

111. Herschel, *Preliminary Discourse*, 1830, p. 134.

112. Herschel to Thomas Maclear, 23 January 1834, RS:HS 21.156. This was about a week after Herschel arrived at the Cape.

113. Sir John Bell, Colonial Office (Cape Town) to Herschel, 18 April 1834, RS:HS 4.18.

114. Herschel to Whewell, 21 September 1834, RS:HS 25.3.21.

115. [John Herschel], *Instructions for Making and Registering Meteorological Observations in Southern Africa, and Other Countries in the South Seas, as also at Sea* (London: Bradbury and Evans, 1835), 17 pp. The original was printed in Cape Town by G. Greig. The London edition was a private printing of 1000 copies, funded by an anonymous donor.

116. Herschel, *Instructions*, 1835, p. 1.

117. Herschel, *Instructions*, 1835, p. 17.

118. Herschel to Hudson, 8 January 1835, *Athenaeum*, 25 April 1835, no. 391, p. 320. Extracts of the *Instructions* were published as: John Herschel, "Instructions for Making and Registering Meteorological Observations, &c." *Athenaeum* no. 401 (4 July 1835): 509–510.

119. Herschel to John Augustus Lloyd, 12 March 1835, RS Sa.645. This correspondence lasted until 1837. Lloyd undertook tidal, pendulum, and geomagnetic

measurements. Lloyd was in charge of the Mauritius observatory of the Colonial Office, where he was mainly charged with calibration of chronometers.

120. Herschel to James Prinsep, 2 December 1834, RS:HS 25.4.4. A dozen letters between Herschel and Prinsep started when Herschel read an article in an Indian journal that confirmed Herschel's ideas on atmospheric circulation. Only two of Herschel's letters survive. Also, Herschel to Edward Ryan, 9 October 1834, TxU:H/L-0765.2, Reel 1089; and Ryan to Herschel, 4 November 1834, RS:HS 14.452. Ryan was then in Calcutta, requesting pointers on reading instruments.

121. The first of seven reports of meteorological results between 1836 and 1838, was: John Herschel, "Hourly Meteorological Observations for the Summer Solstice, Made at . . . Cape of Good Hope, on the 21st and 22nd of December, 1835," *Athenaeum* no. 446 (14 May 1836): 345. He published committee reports: [John Herschel], *Report of the Meteorological Committee of the South African Literary and Philosophical Institution, Read July A.D. 1836* (Cape Town: the Institution, 1836); and Herschel, "Second Report of the Meteorological Committee of the South African Literary and Scientific Institution," *The Edinburgh New Philosophical Journal* 21 (1836): 239–246, on p. 240.

122. Herschel to Beaufort, 11 October 1835, RS:HS 3.343 and (copy) RS:HS 21.188.

123. Jack Morrell and Arnold Thackray, *Gentlemen of Science: Early Years of the British Association for the Advancement of Science* (New York: Clarendon Press, 1981), pp. 353–370; John Cawood, "The Magnetic Crusade: Science and Politics in Early Victorian England," *Isis* 70 (1979): 493–518; and Gregory A. Good, "The Study of Geomagnetism in the Late 19th Century," *EOS* 69(16) (1988): 218–228.

124. Herschel, "Report of the Committee of Physics, including Meteorology," 1840, pp. 53–80.

125. Herschel, "[Presidential] Address]," *British Association Report for 1845* (1846): xxvii–xliv; reprinted as Herschel, "An Address to the British Association," *Essays from the Edinburgh and Quarterly Reviews*, (1857), pp. 652–653.

126. "The Herschel Dinner," *Athenaeum* no. 555 (16 June 1838): 423–427, on p. 427.

CHAPTER 3

Mapping Meteorology

KATHARINE ANDERSON*

Mapping the atmosphere was, and remains, a critical feature of modern meteorological science.[1] As new institutions and practices emerged in the late eighteenth and nineteenth centuries, maps became increasingly important. Through experimentation with the forms and conventions of the weather map, scientists and their audiences showed how they understood these practices and the place of science in modern society. By the end of the nineteenth century, a standard genealogy had already emerged among meteorologists, crystallizing their picture of the development of their science. The influence of these standard geneaologies have made the conventions of mapping, and the corresponding account of the science they upheld, seem much more stable than they in fact were.

In 1897, for example, Gustav Hellman laid out a typical sequence of maps as milestones in the study of meteorology. He reprinted Edmund Halley's map showing the circulation of the trade winds, which appeared in 1688; a map of global isotherms by Alexander von Humboldt, first published in 1817; the synoptic maps of an 1842 storm that Elias Loomis published in the *Transactions of American Philosophical Society* in 1846; a telegraphic weather map from the Paris Observatory in 1863; and an isobar map of average pressures of France from 1864.[2] Hellman limited himself to five examples spread over one hundred and fifty years, but similar collections would usually include mention of a few other key maps, like those of Brandes and Dove. In 1816, Heinrich Brandes had proposed that observations collected by the Meteorological Society of the Palatinate between 1781 to 1792 could be presented in the form of a series of daily maps of Europe,

69

while Heinrich Wilhelm Dove, the Prussian meteorologist and a former student of Brandes', followed his lead in the 1830s and 1840s with maps of temperature averages that displayed a global collection of observations.[3] Maps in the spirit of Humboldt, Brandes and Dove gained further influence and much wider audiences as they were distributed in the physical atlases developed by Heinrich Berghaus (1836) and others.[4]

By the 1870s, when the central meteorological organization in the United States, the U.S. Army Signal Service, began to produce daily maps presenting synoptic observations for North America, the Atlantic and the European continent, the form of the meteorological map seems settled. With their brightly coloured isotherms and isobars sweeping across the space of the map, these international maps are immediately familiar.[5] The standard, significant features of the weather map had become, firstly, scale, which emphasized the need for a global account of the atmosphere; and, secondly, the technique of iso-lines. The iso-lines visually collected and displaced individual numbers—either as a synoptic record, which combined observations taken at the same time in different places, or as a record of averages, which offered a different sort of summary of multiple observations. As Humboldt put it, such maps provided a way to manage the abundance of detail of an observational science, recording a great number of facts in a form that "speak[s] to the senses without fatiguing the mind."[6]

Yet a sustained conviction about the value of maps in identifying meteorology as a modern science of observation did not mean that the scientific community had settled amongst themselves which representational methods were most successful. The nineteenth century, in the wake of both new printing technologies, like color printing, and communication technologies, like the telegraph, saw widespread experimentation with the visual presentation of scientific information.[7] In 1885, George James Symons, who supervised a large network of rainfall observations in Britain from 1860 to 1900, published a description of nearly two dozen possible methods of representing rainfall.[8] A reader leafing through the maps reprinted by Hugo Hildebrandsson and Léon Teisserenc de Bort in their *Bases de la meteorologie dynamique* will see a similar variety of familiar and unfamiliar forms.[9] By 1864, one review of French maps complained that the maps produced by the Paris Imperial Observatory showed only the "audacious dogmatism too common among meteorologists" because their representation of atmospheric change was intelligible only if "learned by heart."[10] In other words, meteorologists were almost universally enthusiastic about mapping, but the conventions of maps themselves were far from fixed. Such diversity, with its accompanying arguments about intelligibility, suggests how maps could provide points of divergence rather than convergence for the discipline of meteorology. While it is certainly possible, then, to glance through the work of nineteenth-century meteorologists and trace a lineage for the standard features of modern maps, it is still more revealing to look at the variety of weather maps, and to consider the debates that accompanied them.

The conventions of maps offer a way to understand the evolution of meteorology into a science of vast collections of observations, a science that required coordination on an international scale, with funding to match. But at the same time, a history of mapping offers us a way to explore the implications of this characterization. Maps were grounded in numbers and the exchange of numbers, and assumptions about a need to build up a picture of global forces, as outlined by influential natural philosophers like Humboldt. As a visual summary of many individual observations, they embodied the spirit of new collective organizations, like the British Association for the Advancement of Science, founded in 1830, or the international magnetic and meteorological observation program of the same decade, led by John Herschel, Edward Sabine and others.[11] Defining meteorology as a global science of massed observation presented a particular account of the relationship between local and global conditions, between the expert and the observer, between the national office and the reader of weather charts. None of those relationships were transparent and simple, and many were deeply political, leading into debates about inductive or empirical methods, public funding for science, the value of pure research rather than practical forecasting, and the accessibility of scientific knowledge.[12]

NATURAL ORDER AND THE SYNOPTIC MAP

Our focus in this chapter will be Brandes' innovation, the synoptic map, which became the map most closely associated with the challenge and promise of weather prediction. As telegraphic weather observation networks took shape in mid-century and synchronized data became easier to obtain, these maps, presenting pictures of the atmosphere at a given moment, became a characteristic part of meteorological work. In 1848, a London newspaper, the *Daily News*, began to print the state of the weather taken at 9 A.M. the previous day at about 30 towns in England and Scotland and forwarded to London by rail. The enterprise was organized in large part by James Glaisher, the head of the meteorological observations at Greenwich Observatory.

Almost from the beginning, these observations were being laid onto skeleton maps at Greenwich for Glaisher and the Astronomer-Royal George Airy to study.[13] Very similar experiments by Joseph Henry and James Espy were taking place at the Smithsonian in Washington, D.C. from 1847 to 1850.[14] In Britain, the earliest published version of the maps came in 1851, at the Great Exhibition in London. There, weather observations, now sent by telegraph, were posted daily on a large map of Great Britain, and penny lithographed versions were printed for visitors to take away.[15] The Smithsonian similarly displayed the telegraphed observations from its network of stations to the public from 1856 on, and newspapers published daily reports (although not maps) shortly thereafter.[16] We can see in these arrangements in the United States and Britain that both the scientific interest of the synoptic map and its popular appeal were immediately recognized by meteorologists.

In particular, constructing a synoptic storm map gave scientific men of this era the powerful psychological experience of discovering natural order in a clutter of details. As Ralph Abercromby, an English meteorologist noted for his work on weather charts and isobaric patterns, claimed, "maps entirely alter the attitude of mind with which we regard weather changes."[17] For Abercromby and many others, the storm map was a compelling object, although the lack of established conventions often made it far more persuasive for its creator than for any other reader, puzzling anew over fresh symbols and notations. One of the best examples here is the work of Robert FitzRoy on the Royal Charter storm of 1859. FitzRoy was the first director of the meteorological department of the Board of Trade in Britain, an office set up in 1854 in response to the conference calling for the coordination of marine meteorological observations in Brussels the previous year. As the routine of collecting meteorological data from merchant marine and naval vessels became established, FitzRoy became more and more interested in building a telegraphic network to forecast storms. His work on the Royal Charter storm of October 24 and 25, 1859 was a critical motivation for these initiatives. Because this dramatic disturbance passed over the middle of England, affecting many densely populated areas, FitzRoy's data was especially extensive. He produced a busy map of isobaric and isothermic lines, wind lines which showed the direction of the wind and indicated its force by their length (longer line, stronger wind), and symbols for cloud cover and different forms of precipitation (Figure 1).[18]

FitzRoy was able to conclude that many telling indications preceded the disturbance, like a very low barometric pressure, a clear atmosphere and magnetic disturbances affecting telegraphic wires. His maps also convinced

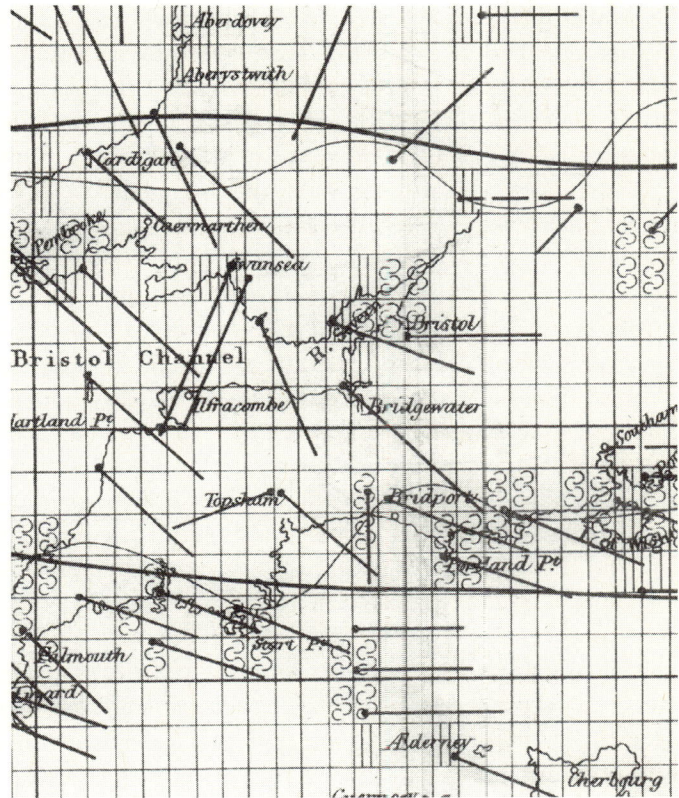

FIGURE 1   Detail of Robert FitzRoy's synoptic map of the Royal Charter storm of 1859, showing observations collected near Bristol and London. FitzRoy, *Weather Book* (1863), appendix.

him to view the atmosphere in terms of contesting currents of warm and cold air in which local eddies, carried along in larger bodies of moving air, produced storm conditions. It was a version of the convective or thermal theories of cyclones, similar to others that were emerging among U.S. and European meteorologists, and it was particularly indebted to Dove's insistence that atmospheric changes depended on the alteration of polar and equatorial currents.[19] As with the Royal Charter storm, FitzRoy translated this theoretical understanding into a visual one: his *Weather Book* of 1862 included the storm map and a hand-colored blue and red representation of the two atmospheric currents.[20] FitzRoy's contribution to debates about cyclone theory was modest at best. Yet maps shaped the essential feature of his meteorological career: his conviction that observations could be carried into the practical arena of forecasting. In that sense, the Royal Charter map was insignificant in comparison to the hundreds of other maps he and his assistant later compiled as a regular part of the work of the meteorological department. These masses of charts, showing "the ordinary course of nature," founded his confidence that weather prediction was not only possible, but "a daily public duty."[21]

FitzRoy's approach to meteorology, in which mapping played such a significant role, is easier to understand when we recall his previous career as an eminent naval hydrographer. In the famous Beagle voyage that transported a young Charles Darwin around the world, FitzRoy's job as captain had been to produce accurate maps of the South American and Pacific coastlines for the Hydrographic Office of the Admiralty. Like his countryman, John Herschel, FitzRoy was fond of the metaphor of atmospheric 'oceans' and spoke of barometric readings as the equivalent of taking soundings.[22] His background in hydrography demonstrates the contemporary understanding of maps as practical as well as scientific objects. Meteorological maps therefore associated the science with two different sets of interests. On the one hand, maps aligned meteorology with the practices of modern science. On the other hand, an emphasis on mapping the atmosphere underlined the relationship of the science to the practical concerns and interests of navigators. In meteorology as in other subjects, scientific men treated the relationship of their science to practical work warily. To John Herschel, writing in 1830, there was no more deplorable question than "cui bono: to what practical end and advantage do your researches tend?"[23] Practical advantages might stimulate interest in science, but they dragged it away from a philosopher's ideals.

Maps and charts, as objects in an exchange between meteorological offices and their networks of observers, were the tangible symbols of a directly utilitarian view of the science of meteorology. As the science of meteorology sought to build networks and establish its research observatories and to control public expectations of forecasting in the 1860s and 1870s, the question of the practical value of the work was never far from the surface of both public debate and private scientific discussion. The individual who best represented this aspect of meteorological mapping was Matthew Fontaine Maury, a lieutenant in the United States navy, and officer in charge of the Depot of Charts and Instruments, a branch of the department of hydrography from 1842 to 1860. Maury had been the driving force behind the international meeting at Brussels in 1853 that sought to coordinate meteorological observations, and for Maury, the science of meteorology was knowledge for the service of navigation, and hence best represented in charts and maps meaningful to navigators. As the collection of marine observations got underway in Europe and the United States and logs started flowing in to central national offices, copies of Maury's published charts and maps were the official recompense that flowed back out to the diligent officers in their "floating observatories."[24] Maury produced charts of preferred sea routes for sailing ships based on collected wind observations that offered to dramatically reduce the duration of voyages, especially to South America, to the west coast of America and Australia. But Maury was also distinctly unpopular with the men pursuing meteorological work at the Smithsonian and elsewhere in the United States, like Joseph Henry and James Espy. Henry and Espy saw him as unqualified and opportunistic, and disliked the speculative nature of his *Physical Geography of the Sea*.[25] Maury's difficult reputation shows how from the beginning of organized national meteorology in the mid-nineteenth-century, the connection of meteorology with the interests of navigation could represent a liability rather than an asset. Maps had a double guise: lowbrow foray into practical science; or highbrow expression of collective, precise scientific observation.

"A WIDE ROOM FOR FANCY": READING THE BRITISH WEATHER MAP

The absence of settled conventions in mapping made any such debate about the value or direction of meteorological work more difficult to settle. Criticism of the charts and maps of FitzRoy's meteorological department in Britain, or Maury's charts in the United States, might have emerged because of the concerns in the scientific community that practical interests were pushing meteorology in the direction of hasty, fragmentary work, seduced by the promise of weather prediction. But on the surface, criticism of the charts was about their confusing appearance. Maury's wind charts were considered in British circles "so puzzling and intricate as to be scarcely intelligible to an ordinary navigator" and "overlaid with meshes of interlacing lines in extraordinary number, so as to resemble entangled skeins of many coloured-threads."[26] FitzRoy's experience in Britain was similar. In the discussions establishing weather telegraphy in Britain in 1860, or during subsequent investigations of the weather office in 1866 and 1877, his storm map of

1859—so decisive for FitzRoy himself—was rarely mentioned. Its notations, at least as intricate as Maury's, made it a puzzling and difficult document to read.

As it became obvious that there were a variety of ways to represent the weather, and that not all meteorologists could read each other's productions with ease and approval, it became correspondingly harder to define the authority of the official meteorologist. Above all, then, maps came to pose questions about expertise. Maps promised meaning at a glance, yet how could the notorious difficulty of meteorology be squared with meaning so transparent that "anyone" could see it? Francis Galton, author of an 1863 collection of weather maps, uneasily recognized the "wide room for fancy" when observations were inadequate. "Exercising the right of occasional suppression and slight modification, it is absurd to see how plastic a limited number of observations become, in the hands of men with preconceived ideas."[27] Moreover, the new networks of telegraphic weather data opened forecasting up to all—that is, to those with preconceived ideas or none at all. With a newspaper in hand, a local barometer and a look outside, any individual could become a kind of weather center himself. How, then, to communicate the training and experience that lay behind the construction and the interpretation of the weather charts? Such concerns surfaced repeatedly in the 1860s and 1870s.[28]

But if encountering the map as a reader, trained or not, was complicated, the role of telegraphy in meteorological mapping in this era nevertheless emphasized the authority of the central office, and the importance of modern, institutionalized science. FitzRoy's telegraphic network centered on his London office, just as the American equivalent centered on the Smithsonian, or the French on the Paris Observatory. In the early 1850s, Maury sought to gather support for an international meeting to coordinate meteorological exchange by emphasizing the confusing variety of on-going land and sea observations, run by states, nations, private institutions or even individuals. But as telegraphy spread over the next decade, the force of such arguments were eroded. Daily telegraphic data networks were so expensive and required so much co-ordination that private equivalents were really not an option. This was one of the main arguments for government leadership—rapid, international collection of meteorological data could not be undertaken by private scientific enterprise. The intervention of an expensive technology like the telegraph created privileged access and control as well as participation and dissemination. The new technique of telegraphy, distributed in the press, may have allowed any observer or newspaper reader to see beyond the confines of their own horizons, but it could also give central offices, as the direct recipient of nation-wide observations, a form of omniscience not available to anyone else.

Several developments in mapping during the 1860s and 1870s responded to these concerns about practitioners and audiences of meteorological science. In March 1872, spurred by the American and French examples, the British Meteorological Office began to produce daily European weather charts. This publication consisted of one sheet, holding four separate maps which recorded isobars, isotherms, wind and sea, and cloud and precipitation observations, with brief remarks underneath each map. The office printed off about 200 copies, by lithography; it was posted outside their office in London and made available to the public by subscription. Within a few years, these charts began to reach a far wider audience, as the London *Times* pressed the Office to supply them for newspaper publication. Beginning in April 1st, 1875, then, the Met Office began to send out daily weather charts and weekly barographic traces that were published in the *Times* and other newspapers.[29]

This 1872 subscription service represented a cautious return to a higher public profile for the Met Office. For the previous several years, since the death of Robert FitzRoy in 1865, the British office had avoided weather predictions and resisted the public pressure for a practical approach to meteorology. The return to daily charts indicated a guarded understanding of the popular interest in meteorology. The weekly scientific journal *Nature* hoped that the charts would provide "a means of leading the public to gain some idea of the laws which govern our weather changes. As soon as they appear in our afternoon papers we may hope for a more intelligent comprehension of the difficulties which beset any attempt to foretell the weather of these islands for the space of even twenty-four hours."[30] According to this account, then, the maps were valuable because they emphasized how difficult it was to understand and predict the weather. The *Nature* article went on to describe in detail the impressive printing technology behind the charts, from the pantographs which reduced the outline map, to the engraving block that could be engraved rapidly "without blurring or chipping," to the templates which would provide a uniform typeface of the characters.[31] The description emphasized the pressure and speed of the process, highlighting the technological foundations of the privileges of the central office. No other observatory or society, let alone an individual meteorologist, had the resources to make these daily charts. By shifting readers' attention to the *process* of image-production, rather than the predictions themselves, the new charts re-asserted the powerful position of the British central office and cautioned the public against easy interpretations of modern scientific data.

In the face of these pressures stemming from the public and practical context of the weather map, its conventional features take on a different significance. By the end of the 1870s, the larger-scale map visually dominated by iso-lines rather than observation points became the more standard, familiar form. Such maps made an argument that the critical work of meteorology would be that work which was controlled by the experts in the central observatories or national offices. Insisting that individual observations were to be subsumed under a general and collective picture of the atmosphere, meteorological enterprises argued that local experience assumed significance only when treated collectively. Maps which emphasized the global scale of the science simultaneously made the individual observer less visible and made the question of local conditions less important.[32]

SCALE, DYNAMISM AND LOCAL OBSERVATION: SHIFTING FEATURES OF BRITISH, FRENCH AND AMERICAN MAPS 1860S–1880S

These patterns can be traced in the examples from Britain, France and the United States that we will consider briefly here (Figures 2 to 4). An experiment in meteorological mapping carried by Francis Galton provides a starting point. Galton, better known from later work in statistics and eugenics, himself came to meteorology from an interest in geographical maps, cultivated during his explorations of south-west Africa in 1850 to 1852. In 1863, he turned his attention to the representation of the weather. His *Meteorographica* presented a series of 93 maps of European weather in December 1861, three per day for the month. It led Galton to a pioneering description of an 'anti-cyclone' or a region of high pressure surrounded in the northern hemisphere by winds moving counter-clockwise.[33] Beyond the value of his project for theories of atmospheric circulation, however, *Meteorographica* represents an unusually explicit statement about the intent of mapping in meteorology. In his preface, Galton argued that maps were at the top of a scientific visual hierarchy. "When observations are printed in line and column, they are in too crude a state for employment in weather investigations; after their contents have been sorted into Charts, it becomes possible to comprehend them. But it requires meteorographic maps to make their meaning apparent at a glance. . . . A few judicious sweeps and shadings of a draughtsman's pen may embody the simultaneous observations of hundreds of meteorologists."[34] The presentation of maps in *Meteorographica*, and Galton's expressed concern to erase "local deviations,"[35] show how the conventions of the weather map evolved.

Galton's maps adopted the technique of color-printing and a particular sequence of maps to emphasize how his maps moved from local observations to global science. Galton's maps were rendered in shades of red and black. Deliberately harsh tints in color printing in cartography, like those Galton chose, were a marked change from subtler hand-colored productions of a previous era. In this early period of color printing, colors were chosen to maximize the effect of the contrast, and to avoid any confusion between similar colors.[36] Stronger colors were, however, also more opaque. That opaqueness had a significant effect on the nature of the representation, moving away from the tradition of an aerial view, towards a more spare, and abstract image. In geographical maps, this became a controversial matter. As the German cartographer Josef von Hauslab protested, "maps have become mere schematic representations instead of pictures."[37] Whatever the implications of these developments for other kinds of maps, when we look at Galton's series of maps in 1863, there seems no doubt that an increasingly schematic representation was exactly the effect he sought.

Galton designed his book to be viewed as a two-day sequence (Figure 2). On opening the book, the reader would see six maps on the verso page and the first half of the recto page, three maps each day with morning, afternoon and evening observations. Those maps held cloud, precipitation, barometer and thermometer readings and wind direction and force, positioned in rectangular blocks on the map. The rest of the recto page held six "small symbolic charts" giving the deviation from a barometric average of 29.95 inches represented in each map, with black denoting lower than the average and red higher, with both the intensity of the shade and design of symbols corresponding to a greater or lesser deviation from the average. The end result was a set of six grids with a checkerboard appearance and, most strikingly, no remaining trace of the outline of the land.[38]

The effect of this sequence, in which scale and design leads the reader in the direction of increasingly abstract conceptions, is even more striking when juxtaposed with Galton's design for a weather map of the British Isles a few years earlier, in 1861. That map showed an outline of the British Isles seen from above, the conventional bird's eye view, including topographical markings. Circular stamps were dotted within the land outline and, by shapes, coloring and numbers marked within the stamps, they encoded wind,

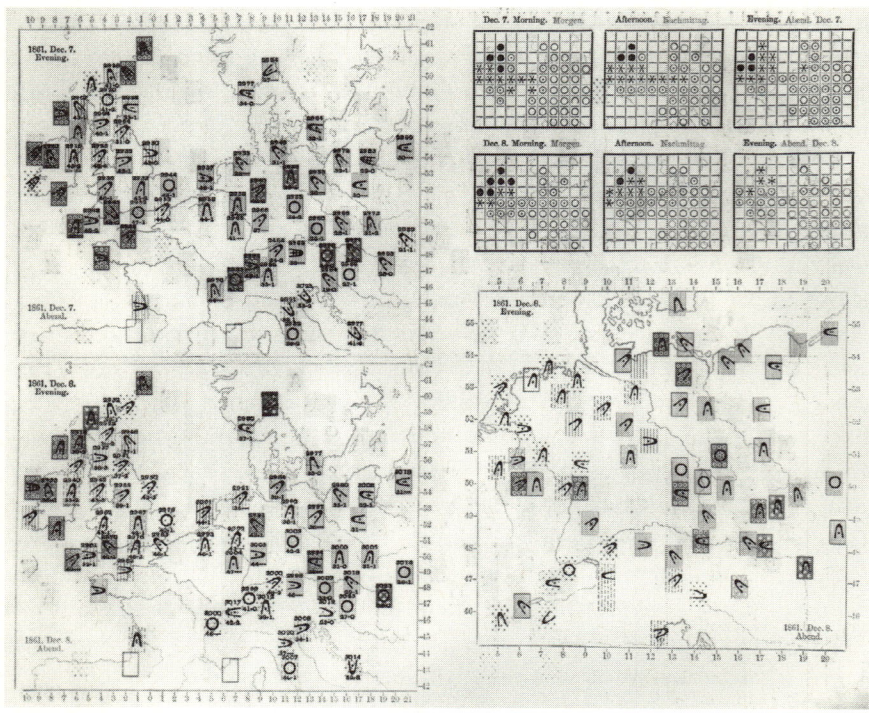

FIGURE 2   Galton's set of maps analyzing weather conditions of December, 1861, provided a sequence of six synoptic charts of western Europe over two days on each page spread, plus summary charts that gave the weather map an increasingly abstract appearance. The verso, shown here, gave the two evening observations for December 8 and 9 (the morning and mid-day observation lay opposite on the previous page) and the accompanying summary charts. Galton, *Meteorographica* (London, 1863).

barometer and cloud information. The stamps were similar to the later mapping techniques in that they held comparable amounts of information, but they were different in how they were positioned and named by specific location on the map.[39] In style, the map was reminiscent of the lithographed daily weather map posted in the Great Exhibition hall in 1851. Considered as a pair, Galton's maps of 1861 and 1863 show a shift from a bird's eye view of separate local observations to a highly schematic vision of the forces of the atmosphere, detached from the observer and point of observation.[40]

The role of maps in underlining a shift away from local accounts of meteorological phenomena emerges as well in the work of Urbain LeVerrier and the Paris Observatory. LeVerrier became director of the Paris Imperial Observatory in 1854, several months before a cyclonic storm destroyed the French fleet in the Crimea. Asked by the French government to investigate the storm, LeVerrier concluded in 1855 that the indications as the storm passed over Europe would have made it possible to telegraph a warning to the Crimea. His subsequent proposal for a telegraphic observation network (organized in 1857) and accompanying storm warning service was thus the earliest in Europe, even though the French government rejected the latter as too costly. (The Belgian and British systems thus took actual priority in forecasting in 1859, while LeVerrier had to wait until 1863 for something similar.) During these years, French meteorological work was split between three locations: the Observatory, carried out chiefly by E. H. Marie-Davy under LeVerrier; by the Ministry of Marine; and for a brief time, by a meteorological observatory in Montsouris as well, a short distance from the Imperial Observatory. It was not until after LeVerrier's death, in 1878, that France organized a central meteorological office.[41] In this period, France acquired a reputation for a lack of international cooperation, failing to participate in international meetings or even to record its views on the issues deliberated there.

What the Paris Observatory did produce in these years was maps. Amid these complicated institutional developments, the publication of maps became a critical expression of LeVerrier's committment to the science. From 1857, LeVerrier published a daily sheet of observations he received by telegraph (initially from fourteen French and five foreign locations), which he called the *Bulletin Meteorologique International*. By the following year, the Bulletin included a lithographed map of the atmosphere over Europe. Shortly afterward, LeVerrier began to collect storm reports, asking observers to note the time the storm began and ended, to give the direction of the winds before, during, and following the storm, to provide observations on clouds, thunder and lightning, and on any damage caused by the storm. These observations were compiled into maps by an assistant, E. Fron, and in 1865, an *Atlas des Orages* published a selection of 44 of these storm maps. The storm atlas was followed by a series of synoptic maps titled *Les mouvements generaux de l'atmosphere*, published intermittently until 1870, when political events in France interrupted the work of the Observatory.[42] LeVerrier was dismissed from his position as director of the Paris Observatory, although he returned in 1873 and stayed until his death in 1878. The publication of the international maps passed to Niels Hoffmeyer in Copenhagen from September 1873 to November 1876, then was taken up again in

December 1880 by G. Neumayer in Germany. Writing at the end of the century, the Swedish meteorologist, Hugo Hildebrandsson, and his French colleague, Teisserenc de Bort, argued that these international maps were the most significant productions of this era—as important as and comparable to the efforts to organize a pan-European meteorological institution that could transcend the political, administrative and financial boundaries of national observatories.[43]

Three significant features of the French publications deserve notice, all of which can be seen in the example reproduced by Hildebrandsson and Teisserenc de Bort, the storm map of 7 May 1865 (Figure 3). Most obviously, the atlas emphasized the need for an international focus in meteorology. With the synoptic maps of storms, showing their movement across Europe, French meteorologists insisted on the insignificance of local conditions. Storms are rarely localized phenomena, they wrote, and argued that the maps demonstrated that, while variations in the soil, elevation, or forests have some impact on lower cloud conditions, their influence on the wider phenomena of the atmosphere is generally of little importance. Such wider phenomena move independently of the land beneath ("avec sa vitesse propre et independante des saillies du sol").[45] The distinction between such phenomena and local storms, such as those produced in mountainous regions, was all important: these latter were dismissively labelled "orages erratiques"[46] (irregular storms) not relevant to the understanding of the general laws of the atmosphere. A second related feature of the French maps was their visual presentation of the movement of the atmosphere. Previous synoptic maps, like Galton's morning and evening maps, described above, or the well-known maps of Buys Ballot in 1852, represented this dynamism by printing a succession of maps showing the state of the atmosphere at different points in time. The *Atlas des Orages* in contrast showed lines marking the passage of the storm at hourly or even thirty-minute intervals on a single image. As Figure 3 shows, these markings of 'storm time,' rather than the isobars that curve around them, became the visually dominant feature of the map. The fleche markings for wind in parallel lines graphically drive the storm on (sometimes lending an impression of railroad tracks plowing across Europe). Finally, of significance are the points marking local damage from lightning and hail. Given the emphasis on broad detail elsewhere in the publications, these marks, representing a toppled tree, electrified cow or damaged hayfield, seem incongruous. One possible explanation reflects the number and status of the observers involved in producing the map. It seems likely that this level of information acknowledged the interests and work of the largely

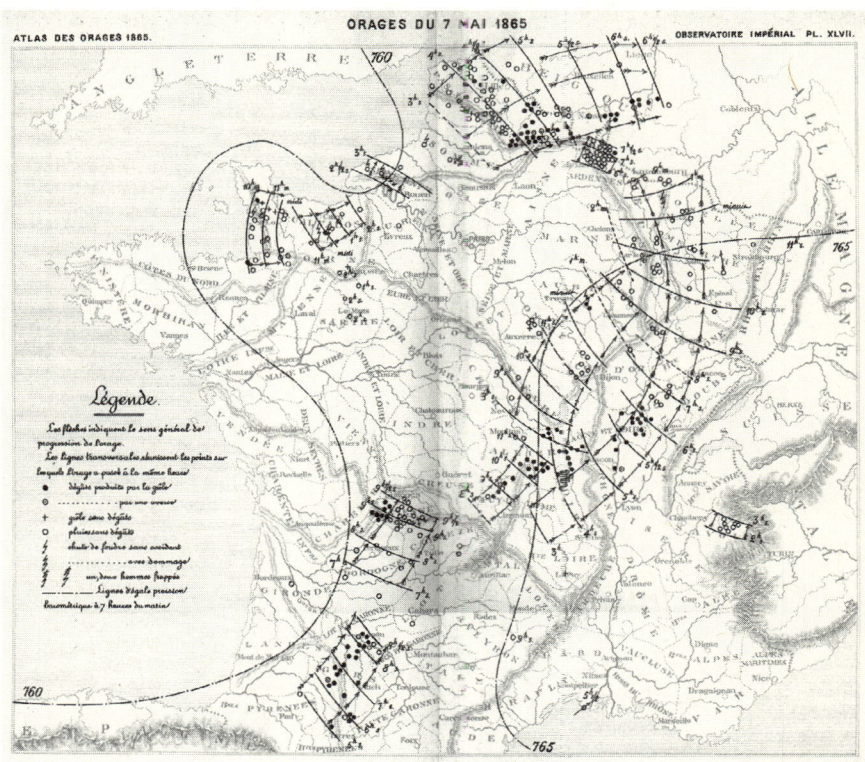

FIGURE 3   Map for 7 May 1865 in *Atlas des orages de l'année 1865*
(Observatoire de Paris, 1866); it was reproduced in H. Hildebrandsson and
L. Teisserenc de Bort, *Les bases de la météorologie dynamique* (Paris: Gauthier-
Villars) 1898, pl. XLVII, opp. p. 244.[44]

volunteer observers, compiling their record for submission to LeVerrier and
the Imperial Observatory in Paris. Whether or not this is indeed the explanation for the combination of local detail and the broad dynamic forces in
this instance, the juxtaposition itself called attention to the continuing need
for meteorologists to seek a balance between a local and global record, if
only as a response to the dependence of meteorology on a large body of committed observers.

A third and final example, the *Bulletin of International Meteorology*
published by the United States Army Signal Service from 1875 to 1884, shows
the confirmation of global scale, centralized and official data collection, and
isolines as the dominant features of the synoptic map (Figure 4).

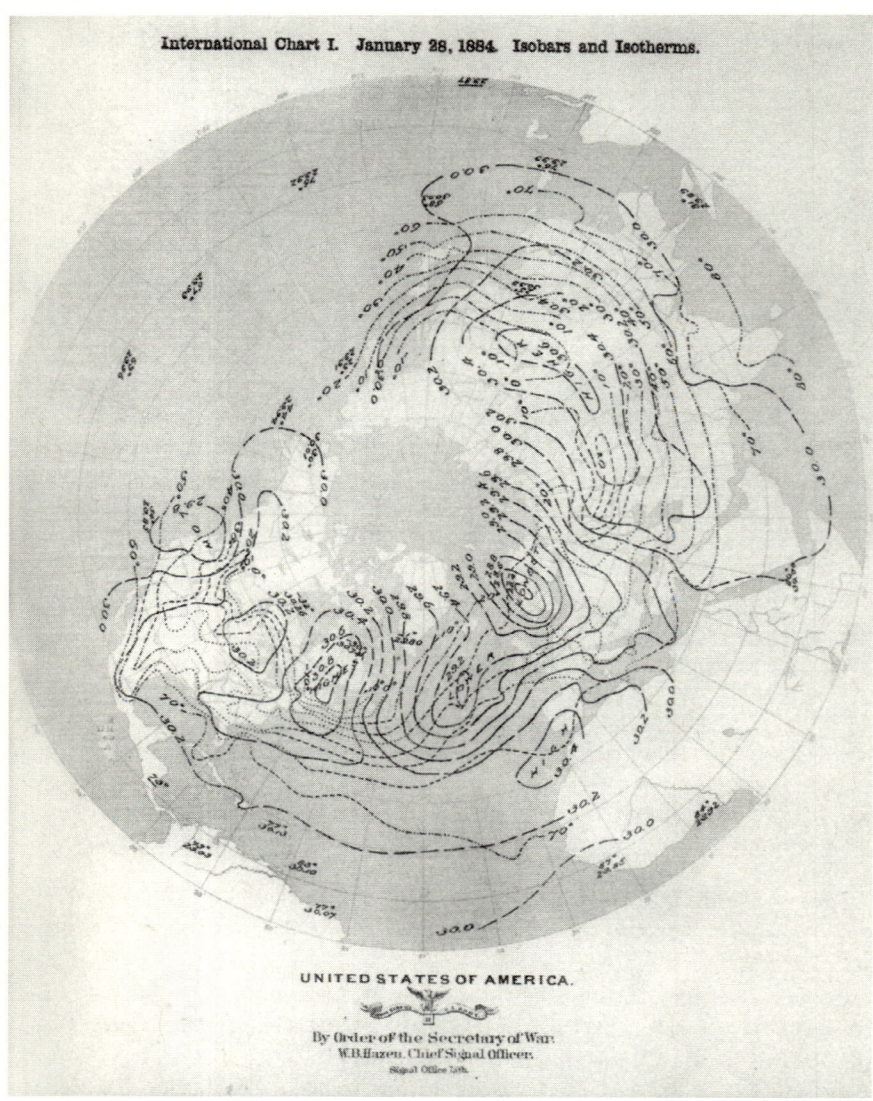

International Chart I.   January 28, 1884.   Isobars and Isotherms.

UNITED STATES OF AMERICA.

By Order of the Secretary of War.
W.B.Hazen, Chief Signal Officer.
Signal Office 1884.

FIGURE 4   By the 1880s, the weather map had taken on an appearance familiar to modern eyes: centered on the north pole, it adopted a global scale, showing isobars as heavier lines, isotherms as dotted lines (original in color). "Simultaneous International Weather Map, 28 January 1884," *Bulletin of International Meteorology* (Washington: Office of the Chief Signal Officer of the Army of the United States, 1884).

The Bulletin published observations from seventeen national centers, listing the director by name at the heading for each section, and, where available, the name of the observers in a column to the far right of the data. The picture given in the Bulletin is of an official, tightly-coordinated global enterprise, staffed by observers who were subordinated to the national leadership of a chief meteorologist. By no coincidence, the best representation of this ideal at this time was the meteorological service of the United States. In 1870, the Smithsonian-based network of Joseph Henry was taken over by a branch of the United States Army, the Signal Service, under General Albert Myer. The remaining volunteer observers of the Smithsonian networks were subordinated to the cadre of trained military observers that Myer could post around the country, and pay for with army salaries.[47] In its scale, expense and centralization, this was an organization without parallel in the European systems. The "simultaneous international weather maps" that began to accompany the Bulletin in late 1877 were thus a predictable outcome of its work. Centered on the North Pole and marking the continents with the spare detail of red isotherms and black isobars, they showed how the global organization of meteorologists matched the global forces of the atmosphere. Questions about particular or local conditions, and about the successful coordination of meteorological observation, still can be heard, but only faintly. Symons, for example, noted that the scale concealed the disorganization of national networks, where observations were often duplicated by institutions that were geographically close but politically and historically distinct, like the two observatories in Greenwich and Kew in London, or the four observatories in Paris. Conversely, the isoline technique concealed the thinness of observations in the Atlantic—with vast areas represented by the observations collected by a few scattered ships.[48] As Symons' comments show, contemporaries acknowledged the implications of the conventional features even as they accepted them.

BALANCING LOCAL RECORDS AND
GENERAL FORCES

The weather map was for meteorologists one of the most prized hallmarks of a modern scientific discipline. It demands, however, a careful reading. While the conventions of the meteorological map can seem at first glance relatively stable, at the end of the nineteenth century this stability was still a recent development. Likewise, the standard genealogy of the weather map, following the lead of late nineteenth-century writers, has glossed over much of the experimentation with its techniques. This is particularly true of the middle decades of the nineteenth century, a vital period when telegraphic weather observation networks and the establishment of national centers for meteorology was transforming the science. The intention here is not so much to construct a evolutionary path that is more sensitive to varieties and local interpretations—although such a history would be interesting. Rather, the goal is to suggest how the experimentation itself can be examined as a way to illuminate key organizational and epistemological issues that faced the science.

During this era, meteorologists sought to develop a balance between the overwhelming detail, even clutter, of particular observations and general dynamical laws. They also tried to find a balance between meteorology as a public science, concerned with weather prediction, and meteorology as a research science. As part of their sense of the scientific importance of maps, then, meteorologists recognized that maps shaped the relationship between elite and popular forms of weather knowledge. Any weather map offered simultaneously a convenient method for summarizing data, a claim about the intelligibility of the atmosphere, and a public object that defined the identity of modern meteorological science. The experimentation with weather maps in the middle decades of the century in the United States, Britain and France reveals a shift away from types of representation that summarized individual data, but retained the record of individual or local detail, towards types of representation that also summarized multiple observations, but subordinated them to broad patterns, using methods that could reveal how both the science and the phenomena transcended a local scale. These conventions visually summarized critical features of meteorology in this period: firstly, meteorologists' insistence on the importance of multiple, extensive observations in order to advance meteorological knowledge; secondly, the pressure to move beyond mere data collection to discussion of physical laws; and thirdly, the need to control public expectations about the practical value of meteorology, particularly weather prediction. Because maps were perhaps the most noted productions of modern government-sponsored meteorological research, moreover, these techniques developed within the context of larger debates about the public value of science. As techniques that reflected the tension between observations and explanation in meteorology, and as public objects whose intelligibility was a matter of open debate, weather maps offer a sometimes surprising picture of meteorology.

NOTES

*This chapter emerged from a presentation at the ICHM's Beaufort to Bjerknes conference in July 2004 and benefited from the engagement and attention of participants of that meeting. I owe special thanks to Debbie Coen, Vlad Jankovic, and Jim Fleming for their comments, suggestions, and patience as well as to staff at the NOAA Central Library, the Environment Canada Library (Downsview), and York's Resource Sharing Department.

1. Other accounts of the development of mapping include Mark Harrington, "History of the Weather Map," *Report of the International Meteorological Congress 1893*, United States Weather Bureau Bulletin no. 11 (Washington D.C., 1894), 327–35; Mark Monmonier, *Air Apparent: How Meteorologists Learned to Map, Predict and Dramatize Weather* (Chicago: University of Chicago Press, 1999); William Napier Shaw, *Manual of Meteorology*. Vol. 1: *Meteorology in History* (Cambridge: Cambridge University Press, 1926), esp. 288–315. See also Arthur H. Robinson and Helen Wallis, *Cartographical innovations: An international handbook of mapping terms to 1900* (Tring, Eng.: Map Collector Publications in association with the International Cartographic Association, 1987) and Arthur H. Robinson and Helen Wallis, "Humboldt's Map of Isothermal Lines: A Milestone in the History of Thematic Cartography," *Cartographic Journal* 2 (1967): 119–23; Arthur Robinson and Barbara Bartz Pentchenik, *The Nature of Maps: Essays towards Understanding Maps and Mapping* (Chicago: University of Chicago Press, 1976).

2. Gustav Hellmann, *Meteorologische karten 1688, 1817, 1846, 1863, 1864. Sechs tafeln in lichtdruck mit einer einleitung* (Berlin: Asher, 1897).

3. In 1853, the Royal Society of London awarded Dove its highest honour, the Copley Medal, for this work, commending Dove's "true spirit of inductive inquiry" which had produced "real order in midst of apparent confusion." Parsons [Earl of Rosse], "Address Delivered Before the Royal Society," *Abstracts of Papers Communicated to the Royal Society*, 6 (1850–54): 353–54.

4. Jane Camerini, "The Physical Atlas of Heinrich Berghaus. In R. G. Mazzolini, ed., *Non-Verbal Communication in Science Prior to 1900* (Florence: Leo S. Olschki), 479–512.

5. *Bulletin of International Meteorology*. (Washington DC: Office of the Chief Signal Officer of the Army of the United States, 1875–1884).

6. Alexander Humboldt, *Political Essay on the Kingdom of New Spain* (London: Longman, 1811), cxxxii, quoted in Robinson and Wallis, "Humboldt's Map," 122. Cf. Michael Dettelbach, "The Face of Nature: Precise Measurement, Mapping and Sensibility in the Work of Alexander von Humboldt," *Studies in the History and Philosophy of Biology and Biomedical Sciences* 30 (1999), 473–504.

7. A key text in understanding these developments is Lorraine Daston and Peter Galison, "The Image of Objectivity," *Representations* 40 (Fall 1992): 81–128. For a survey of the literature, see Alex Pang, "Visual representation and post-constructivist history of science," *Studies in the History and Philosophy of Science* 28 (1999), 139–171. For visual methods other than mapping in meteorology, see Katharine Anderson, "Looking at the Sky: the Visual Context of Victorian Meteorology," *British Journal for the History of Science* 36 (2003), 301–32.

8. George James Symons, *British Rainfall 1884* (London: Edward Stanford, 1885), 24–27.

9. Hugo H. Hildebrandsson and Louis Teisserenc de Bort, *Les bases de la météorologie dynamique* (Paris: Gauthier-Villars, 1898).

10. *Reader* (London), 19 Dec. 1863, 730.

11. John Cawood, "The Magnetic Crusade: Science and Politics in Early Victorian Britain," *Isis* 70 (1979), 493–518; Jack Morrell and Arnold Thackery, *Gentlemen of Science: The Early Years of the British Association for the Advancement of Science* (Oxford: Clarendon, 1982).

12. On nineteenth-century meteorology, see Anderson, *Predicting the Weather: Victorians and the Science of Meteorology* (Chicago: University of Chicago Press, 2005); James Rodger Fleming, *Meteorology in America 1800–1870* (Baltimore: Johns Hopkins University Press, 1990); Charles Bates and John Fuller, *America's Weather Warriors 1814–1985* (College Station: Texas A&M University Press, 1986).

13. William Marriot, "Earliest Telegraphic Meteorological Reports and Weather Maps," *Quarterly Journal of the Royal Meteorological Society* 28 (1902), 123–131.

14. Fleming, *Meteorology in America*.

15. Marriot, "Earliest Telegraphic Meteorological Reports and Weather Maps." Cf. the Daily Weather Map Company, a floated prospectus of 1861, described in William Marriott, "The Bequest of George James Symons," *Quarterly Journal of the Royal Meteorological Society* 27 (1901), 258–59.

16. Fleming, *Meteorology in America*, 141–45; Monmonier, *Air Apparent*, 39–42.

17. Ralph Abercromby, *Weather: A Popular Exposition of the Nature of Weather Changes from Day to Day* (London: Kegan Paul, Trench, 1887), 10.

18. Robert FitzRoy, *The Weather Book: A manual of practical meteorology* (London: Longman and Green, 1863), 298–329, App. XIII; FitzRoy, "On British Storms, Illustrated with Diagrams and Charts," *Report of the British Association for the Advancement of Science 1860: Transactions of the Section* (London: 1861), 39–44. Herbert Lamb, *Historic Storms of the North Seas, British Isles and Northwest Europe* (Cambridge: Cambridge University Press, 1991) 135–36.

19. Gisela Kutzbach, *The Thermal Theory of Cyclones: A History of Meteorological Thought in the Nineteenth Century* (Boston: American Meteorological Society, 1979).

20. FitzRoy, *Weather Book*, App. VI–VII, XIII.

21. FitzRoy, *Weather Book*, 103. On FitzRoy and forecasting, see James Burton, "Robert FitzRoy and the Early History of the Meteorological Office," *British Journal for the History of Science* 19 (1986), 147–76 and Anderson, *Predicting the Weather*, 105–130.

22. H.E.L. Mellersh, *Fitzroy of the Beagle* (London: Rupert Hart Davis, 1968); Duncan Carr Agnew, "Robert FitzRoy and the Myth of the Marsden Square: Translatlantic Rivalries in Early Marine Meteorology," *Notes and Records of the Royal Society of London* 58 (2004), 21–46; John Herschel, *Meteorology* (Edinburgh: A. & C. Black, 1861); cf. the atmospheric wave theories which also drew on the analogy between tides of the atmosphere and ocean, Vladimir Jankovic, "Ideological crests versus empirical troughs: John Herschel's and William Radcliffe Birt's research on atmospheric waves 1843–1850," *British Journal for the History of Science* 31 (1998), 21–40.

23. Herschel, *Preliminary Discourse on the Study of Natural Philosophy* (Reprint ed. Chicago: University of Chicago, 1987). 10. A good introduction to these tensions in the British context is the work of Roy MacLeod, *Government and Expertise: Specialists, Administrators and Professionals, 1860–1900* (Amsterdam: Elsevier Science Publishers, 1980).

24. John Locke, *Meteorology of the Sea* (Dublin, 1859), 4.

25. Stephen Dick, *Sky and Ocean Joined: the U.S. Naval Observatory 1830–2000* (Cambridge: Harvard University Press. 2004); Charles Lee Lewis, *Matthew Fontaine Maury: Pathfinder of Seas* (Annapolis: United States Naval Institute, 1927); Marc I. Pinsel, "The Wind and Current Chart Series Produced by Matthew Fontaine Maury," *Navigation* 28, no. 2 (1981): 123–137. For an account of Maury's maps based on actor-network theory, see T. Hugh Crawford, "Networking the (non)human: 'Moby Dick,' Matthew Fontaine Maury and Bruno Latour," *Configurations* 5 (1997), 1–21. On low regard for Maury amongst Americans see Fleming, *Meteorology in America*, ch. 5.

26. *Report of a Committee to Consider Certain Questions Relating to the Meteorological Department of the Board of Trade* (London: H. M. Stationery Office, 1866), 13n.

27. Galton, *Meteorographica, or Methods of Mapping the Weather* (London, privately printed, 1863), 3–4.

28. FitzRoy, *Report of the Meteorological Department* 1862, 457–58; Anderson, *Predicting the Weather*, 187–210.

29. R. P. W. Lewis, "The Daily Weather Report and associated publications: 1860–1980," *Meteorological Magazine* 111 (1982), 103–119.

30. "The Times Weather Chart," *Nature* 11 (15 April 1875), 474; Robert Scott, "Weather Charts in the Newspapers," *Journal of the Royal Society of Arts* 23 (1875): 776–782.

31. Karl Pearson, *Life, Letters and Labours of Francis Galton*, 3 vols. in 4 (Cambridge: Cambridge University Press, 1914–30), II: 44–47.

32. The history of geographical maps and local observers gives a similar picture. See Christian Licoppe, "The project for a map of Langedoc in eighteenth-century France at the contested intersection between geography and astronomy," in Marie-Noelle Bourguet, Christian Licoppe and Otto Sibum, ed., *Instruments, Travel and Science: Itineraries of Precision from the Seventeenth to the Twentieth Century* (London: Routledge, 2002), 51–74.

33. Pearson, *Life of Galton*, II: 39–40.

34. Galton, *Meteorographica*, 3.

35. Galton, *Meteorographica*, 5.

36. Ulla Ehrensvard, "Color in Cartography: A Historical Survey," in David Woodward, ed., *Art and Cartography: Six Historical Essays* (Chicago: University of Chicago Press, 1987), 123–46.

37. Ibid., 143.

38. Even his sympathetic student and biographer, Karl Pearson, regarded these maps with distaste. Pearson, *Life of Galton*, II: 41.

39. This map is also reprinted in Pearson, *Life of Galton*, II, pl. VI.

40. The convention of the bird's eye view in mapping, and its role in scientific images, deserves further exploration. Mid-nineteenth-century aerial maps of towns, for instance, were sometimes accomplished from ballooning expeditions, which suggests a particular connection of this convention to meteorological work. On the bird's eye view, see David Buissert, ed., *From Sea Charts to Satellite Images: Interpreting North American History through Maps* (Chicago: University of Chicago Press, 1990); on astronomers and mapping in this period, see William J. Ashworth, "John Herschel, George Airy and the Roaming Eye of the State," *History of Science* 36 (1998), 151–76; for the satellite version of the aerial view and meteorology, see Jody Berland, "On Reading the Weather," *Cultural Studies* 8 (1994), 99–114.

41. John L. Davis, "Weather Forecasting and the Development of Meteorological Theory at the Paris Observatory 1853–78," *Annals of Science* 41 (1984), 359–82; Alfred Fierro, *Histoire de la météorologie* (Paris: Denoel, 1991); Fabien Locher, "Le Nombre et le Temps: La météorologie en France (1830–1880)" (Ph.D. diss., EHESS, Paris, 2004).

42. The most accessible account of the French maps is that in Hildebrandsson and Teisserenc de Bort, *Les bases de la météorologie dynamique*. See also the survey of meteorological work in Alfred Grandidier, *Exposition universelle internationale de 1878: Rapport sur les cartes et les appareils de géographie et de cosmographie sure les cartes géologique et sur les ouvrages de météorologie et de*

statistique (Paris: Imprimerie Nationale, 1882). (Note that in the 1878 international exhibition, all work in meteorology was considered under the general class of maps.)

43. Hildebrandsson and Teisserenc de Bort, *Les bases de la météorologie dynamique*, 134–36. For discussions about the international office for meteorology, which were part of the early debates of the meteorological congress formed in 1872, see *Report of the Proceedings of the Meteorological Congress at Vienna: protocols and appendices* (London: H.M. Stationery Office, 1873); *Report to the permanent committee of the first International meteorological congress meeting at Utrecht 1878* (London: H.M. Stationery Office, 1879); Hugo Hildebrandsson and Gustav Hellmann, *Codex of Resolutions adopted at International Meteorological Meetings 1872–1907* (London: H.M. Stationery Office, 1909).

44. I am grateful to Skip Theberge at NOAA Central Library for assistance in obtaining a copy of this map from the original.

45. Hildebrandsson and Teisserenc de Bort, *Les bases de la météorologie dynamique*, 244–45. "Les orages ne sont pas en général des phénomènes localisés. . . . Les cartes montrent que, si les accidents de sol, les chaines de montagnes, les vallées, les forêts exercent une action sur les nuages inférieures, leur influence est le plus souvent peu considérable sur l'ensemble du phénomène, qui se propage dans les régions supérieures, avec sa vitesse proper et indépendante des saillies du sol."

46. Hildebrandsson and Teisserenc de Bort, *Les bases de la météorologie dynamique*, 249.

47. James Rodger Fleming, "Storms, Strikes, and Surveillance: The U.S. Army Signal Office, 1861–1891," *Historical Studies in the Physical and Biological Sciences* 30, no. 2 (2000): 315–32; Fleming, *Meteorology in America*; Bates and Fuller, *Weather Warriors*.

48. "Daily Bulletin of International Meteorological Observations," *Symons' Monthly Meteorological Magazine* 14 (1879), 50–52. Cf. "International Meteorology," *Symons' Monthly Meteorological Magazine* 8 (1873–74), 181–82.

CHAPTER 4

Fog, Dust and Rising Air

Understanding Cloud Formation, Cloud Chambers, and the Role of Meteorology in Cambridge Physics in the Late 19th Century

RICHARD STALEY[*]

INTRODUCTION

In early 1895, the paths of two Cambridge scientists crossed as teacher and post-graduate researcher. Just days apart, Napier Shaw and C.T.R. Wilson both used cloud chambers to pursue common interests in the study of cloud formation, introducing samples of air into glass globes and watching the droplets formed when that air was subjected to a sudden expansion. On the 20th of March, Shaw gave a lecture before the Royal Meteorological Society meeting in London, "On the Motion of Clouds Considered with Reference to Their Mode of Formation." He was then a lecturer in experimental physics at the University of Cambridge, having played a central role in re-forming the pedagogical standards of the Cavendish Laboratory over the previous decade. Shaw had given classes on meteorology and written an exhaustive study of dew-point instruments for the Meteorological Council of the Royal Society. The cloud chamber was just one of several devices he used to illustrate different facets of the weather in his London lecture. By 1900, he had been asked to become secretary (and later, Director) of the Meteorological Office, where he led the expansion of that institution and authored

93

the classic *Manual of Meteorology* as well as the engaging popularization *The Drama of Weather*. He has been described as a father of modern meteorology; but his interest in cloud chambers has drawn little attention.[1]

By contrast, in 1895 Wilson was a demonstrator in physics for medical students at Cambridge. Having finished the Natural Sciences Tripos and then tried school teaching, he had decided to take the next steps towards a career in research physics by coming back to the Cavendish without a secure faculty or college position. Just five days after Shaw spoke, Wilson opened a new section in his laboratory notebook and commenced a series of experiments in which he filtered the air passing into the cloud chamber before measuring the critical expansion ratio at which clouds formed in *dust-free* air. Wilson never left Cambridge again and devoted much of his career to research with the cloud chamber, becoming particularly well known for the photographs of particle tracks that he began making with the device from 1911. In 1927, Wilson received the Nobel Prize in physics "for his method of making the paths of electrically charged particles visible by condensation of vapour."

The cloud chamber Wilson used has a celebrated history. As it is currently told, that history moves decisively between mountaintops and laboratory benches, and mimetic experimentation and analytic physics. Wilson himself described his initial use of the instrument in romantic terms, as inspired by observations of haloes and glories at the summit location of the Ben Nevis Observatory in Scotland.

> In September 1894 I had spent the fortnight working at the Observatory on Ben Nevis which was to lead both to my cloud chamber work and to my lifelong interest in electricity.

Just over six months later in the Cavendish Laboratory, Wilson got to work:

> Now I began to make experiments on clouds, made by the expansion of moist air, attempting to reproduce the beautiful optical phenomena of the coronas and glories I had seen on the mountain top. I almost at once found indications of the existence of nuclei other than Aitken's dust particles, and designed apparatus for studying them.[2]

Historians of science have followed Wilson's lead in developing a more elaborate and epistemologically oriented account. Galison and Assmus argue that while beginning with a device that John Aitken and others had used to mimic the formation of cloud on dust in nature, Wilson's introduction of a cotton-wool filter radically changed the purpose of the instrument. Showing his allegiance to the tradition of analytic physics fostered by J.J. Thomson at the Cavendish, the filter rendered Wilson's cloud chamber an instrument to probe an artificial phenomena: the formation of droplets on *ions* rather than on dust. According to Galison and Assmus, while Wilson's broader research on atmospheric electricity continued to straddle the divide between meteorological concerns and ionic physics, the cloud chamber itself had decisively crossed the border into fabricated phenomena and the analytic aims of an emerging matter physics.[3]

Both accounts overlook the role Shaw might have played in the development of Wilson's research. The first aim of this chapter will be to show in general terms both why Shaw was important in the Cavendish laboratory, and why he has received so little notice from later commentators. My exploration of the conditions of pedagogical and disciplinary visibility in Cambridge will draw out central differences between the nature of Shaw's teaching as a demonstrator, and the role J.J. Thomson assumed as professor and director of the laboratory. It will also document an important phase in the changing relations between physics and meteorology. If familiarity breeds contempt, I will show here that intimacy fosters a selective invisibility. Following the three scientists Shaw, Thomson and Wilson through the corridors and garrets of the Cavendish will indicate various ways in which the intimate relations of a laboratory at work involved the celebration of certain ties, while allowing other (no less formative) bonds to settle into obscurity. Importantly, we will find that despite the fact that Wilson never mentioned Shaw, and even though the two men surely used the cloud chamber in different ways in 1895, they shared a common theory of cloud formation. In the second half of this chapter, I will show that this theory gave a justification to Wilson's experiments that rendered them continuously mimetic, thereby contradicting the disjunctions historians have highlighted: even when in search of elementary foundations, Wilson's experiment remained both mimetic and analytic. Recovering Shaw's work will show that alongside Thomson's matter physics, Cambridge fogs played their part in inspiring a new use of the cloud chamber. Our story will turn on the particular work it takes—intellectual, empirical, and social—to correlate the intimate details of laboratory manipulation with universal elements or natural processes.

PRACTICAL PEDAGOGY: THE GUTS
OF METEOROLOGY, THE GARRETS OF
LABORATORY PHYSICS

William Napier Shaw (1854–1945) came from a family of manufacturing
goldsmiths and jewelers in Birmingham, and went to Cambridge intending
to study for the India Civil Service. His scholarship, however, was meant for
the Mathematics Tripos, so Shaw completed the most prestigious Cambridge
degree before moving on to subjects he had loved at school: chemistry and
physics. Shaw took physics under James Clerk Maxwell just two years after
the new Cavendish Laboratory opened in 1874, gained first class honors in
the Natural Sciences Tripos, and was then offered a fellowship at Emmanuel
College. Over the next 20 years, he developed a research and teaching program that is best described as *practical*. Early on, Shaw specialized in thermodynamics and electrolysis. When the Meteorological Council of the Royal
Society wanted someone to compare methods of determining the humidity
of the air in 1879, Maxwell directed them to Shaw. That research task led
him to give his first college lectures on meteorology. Years later, he wrote
that now such a course would hardly be regarded as "within the pale of academic physics."[4] He also offered a fascinating description of the physical
sciences of the late nineteenth century:

> At that time the physicist and the chemist always drew from the at-
> mosphere the illustrations of their laboratory theories and conclu-
> sions. His duty was to find explanations for the sequence of events
> which had been observed at a given place, and he did.[5]

Despite the importance of the atmosphere, a note in Shaw's archives
headed "The Cambridge Attitude Towards Meteorology" gives a vivid picture of the differentiated organs and low standing of the science of meteorology in his university:

> "My son's had a good eddication he's passed the locals he says you
> don't want to go to no lectures [on Meteorology] if there is anything
> you wants to know I'll learn you myself."

Meteorology is geography or physics or mathematics according to
the problem under investigation. We have a school of geography, a
famous school of physics, and a whole faculty of mathematics. So

of meteorology as a science we have the legs and the lungs and the head and anybody who has got the legs and the lungs and the head has got the whole bird, except the guts, which are too disgusting for scientific gentility.[6]

Meteorology was to become Shaw's primary specialty, but historians of physics know him best for his work with Richard Glazebrook. Together they acted as laboratory demonstrators when Lord Rayleigh took over from Maxwell in 1879, and pioneered laboratory classes for ever-growing numbers of students. They codified their practices in the textbook *Practical Physics*, which went through four editions between 1885 and 1893 and quickly became Britain's most important laboratory teaching text. It specified recipes for the completion of standard experiments, and helped physicists make efficient use of instrumental resources, emphasizing the importance of careful measurement and leading students to the mastery of techniques central in mechanics, hydro-mechanics, heat, acoustics, light, electricity and magnetism.[7]

But their work was controversial within Cambridge itself, a fact that is indicated by the unexpected election of J.J. Thomson over Glazebrook when Rayleigh left in 1884. Thomson was just 27 years old and clearly a far better mathematical physicist than he was experimentalist. The appointment seemed to demonstrate that the old school still held power in Cambridge. Rayleigh was alarmed and wrote urging Shaw to stay for at least a few terms, for the sake of the laboratory.[8] Fortunately for the Cavendish and its new professor, Glazebrook and Shaw did stay, providing a crucial continuity of expertise and assisting Thomson both in teaching experimental physics and negotiating Cambridge politics. Shaw himself consistently showed a practical concern for those on the margins and crossing the borders of the elite institution. He contributed to the organization of intercollegiate open scholarship examinations, supported degrees for women and helped establish an Appointments Board to bring employers into contact with graduates.

Pedagogically, I want to highlight the way Shaw encouraged students to make the transition from instruction to research, in the period in which it became possible to become a professional physicist.[9] His classes began with tight, controlled instruction for undergraduates. Then the relations between students, apparatus and content were loosened, by staging collaborative and competitive classroom work with a variety of instruments. Students would be given similar tasks using different materials and instruments of varied caliber (Shaw would sometimes test more mature students by giving them defective apparatus). Pooling their experience would help the class build a broad knowledge base. Finally, advanced research students were encouraged to take an independent approach to original memoirs. Both Shaw and Glazebrook offered courses in their specialties, with Shaw teaching "Thermodynamics and Radiation" in the Lent term and "Electrolysis" in the Easter terms of 1888 to 1890, and Glazebrook offering courses on "Electrical Measurements" every Lent term from 1890. Interestingly, H.F. Newall (Shaw's successor as demonstrator) reported that in their advanced classes the different aims of the *students* made the combined form of teaching impossible. He described the essence of Glazebrook and Shaw's approach in this context as being to refer students to original memoirs, commenting "there can be but little doubt that this method leads best to the easy development, not only of the instinct of research, but also of individuality in the investigator."[10] Glazebrook and Shaw published the introductory text *Practical Physics* under their own names. The year after it appeared, Shaw *edited* a small volume of studies in heat that stemmed from his advanced class of 1884. It contained the signed contributions of nine students who had recapitulated classic experiments in thermodynamics. It is worth noting also that meteorological interests were reflected in experiments on the specific gravity of a vapor, measurement of the saturation tension of water vapor in vacuo and in air, and a dew-point measurement verifying Regnault's formula for wet and dry bulb thermometers.[11] Recognizing the careful attention Shaw paid to cultivating the students' own responsibility, I want to suggest that in contrast to the professorial role J.J. Thomson assumed, Shaw may have trained his own disappearance into the approach students like Wilson subsequently took to research. In the second part of this chapter, a study of Wilson's notebook will indicate revealing differences in the relative visibility of Wilson's teachers, even in the pages of his notebook.

Ironically, pointing to Shaw in Cambridge is a way of drawing attention to a kind of physics that on the one hand became part of the furniture, and on the other hand went beyond its walls. His pedagogical work with Glazebrook was absorbed into the structure of the laboratory. As the two men received positions commensurate with their level of responsibility under the younger and less experienced professor, their demonstratorships were taken over by men they had trained (and of course their techniques were propagated still more widely through *Practical Physics*). But as we shall see below, their research programs linked the Cavendish to other institutions; and were gradually displaced from the laboratory to find new institutional homes around the turn of the century. Glazebrook became the first Director of the National Physical Laboratory in 1898 and Shaw was first Secretary and then Director of the Meteorological Office from 1900. In his obituary of Shaw, E.E. Gold described a group of physicists as striving to "weld the geophysical sciences into a coherent whole and prevent their relegation to a garret in the mansion of laboratory physics." As well as Shaw, he was thinking of Glazebrook, Arthur Schuster, H.H. Turner, George Darwin and C.V. Boys.[12]

Already we have seen enough to suggest that meteorology was an increasingly liminal facet of physics in the late nineteenth century, with a substantial shift in status registered in the very period in which the physics discipline became strongly established as a laboratory science. Robert Kohler has argued that in a similar period, the rise of laboratory biology decisively altered perceptions of different features of what had been a broadly shared set of tools, with a consequent devaluation of field studies; and has urged historians to consider the laboratory-field relationship as a cultural geography by focusing on the concept of boundary objects or, better, borderland zones.[13] Galison's analysis of Wilson offers a great deal of support, interpreting his broad concerns with different facets of condensation physics between 1895 and 1911 as cutting across a rich constellation of interests that bridged meteorology, climatology, the physics of steam engines, and analytic matter theory. But Shaw's career suggests the value of taking Kohler's approach to the relations between meteorology and physics even more generally: for Wilson was far from an isolated individual in the stronghold of matter physics. Rather, Wilson's meteorological interests were cultivated within the Cavendish itself. Similarly, I will argue here that it is only in retrospect that Wilson's cloud chamber has appeared to stand as a decisive crossroad between divergent interests.[14]

Kohler regards instruments as significant features of the border zones between laboratory and field sciences generally, often indicating a great deal about the cultural dynamics of communities in flux. Here we will see that instruments provided fundamental but heterogeneous points of contact for those working between physics and meteorology. They served as devices recapturing indoors a mimetic representation of heroic, mountain-top observations like glories.[15] They enabled the measurement of important parameters outside the laboratory like the dust content of air, dew points and temperature (and Kohler has shown that meteorological instrumentation was to be critical for ecological sciences through the twentieth century).[16]

Shaw and Wilson's work will indicate that instruments also provided illustrative models of meteorological phenomena for laboratory pedagogy and, as quantitative investigative tools, offered the foundation for predictive arguments about those phenomena. Ultimately, disputes about the cloud chamber involved settling issues about the borders between artificial and natural phenomena, but had to reach far beyond glass and filters to do so.

PRACTICAL RESEARCH AND SPECULATIVE METEOROLOGY: NAPIER SHAW ON "THE MOTION OF CLOUDS CONSIDERED WITH REFERENCE TO THEIR MODE OF FORMATION"

Shaw's first research for the Meteorological Council involved determining the efficiency and limits of available "dew point" instruments and hygrometers.[17] Shaw determined the amount of water vapor present in both saturated and unsaturated air by extracting it in drying tubes. Comparing current instrumentation, Shaw concluded that no practical methods of observation enabled the water vapor in the atmosphere to be determined within one percent. In the field of meteorology, this was good standards work. Determining standards held a particular importance for the relatively new institutions of academic physical laboratories. Standards provided a secure basis for theoretical and instrumental advance, but were also central to establishing relations between physics laboratories and observatories, and the practical concerns of the nation and its commerce. It was to a large extent through standards that physicists sought to enter the borderlands between their field, other disciplines, and industry. Maxwell had earlier played a key role in the British Association Committee on electrical standards, undertaking experiments designed to install the Ohm as an international standard of resistance. Rayleigh, Glazebrook and Shaw all joined that committee when it was reconstituted from 1881, setting up a network of standards that linked the Cavendish Laboratory to the routine measurement of resistance throughout the electrical industry.[18] As a member of the British Association Committee on electrolysis, in 1890 Shaw prepared a comprehensive review of research on electrolysis and electrochemistry.[19] Its aim was to enable readers to form an opinion of the extent to which electro-chemical phenomena had been referred to mechanical processes and mechanical or electromechanical laws, and to give indications of the probable directions of future progress. Shaw clearly specialized in the report form of research, offering a synthetic review of the instruments, experiments and theoretical contributions to a particular field.

But the scientist displayed a more speculative register when asked to lecture at the Royal Meteorological Society in London on the 20th of March 1895. He chose to speak on the motion of clouds and their mode of formation. To this audience, Shaw described himself as a "mere pedagogue in Physics," and emphasized how much the ideal atmosphere of the laboratory or lecture room differed from actual meteorological phenomena.[20] His lecture in London offers the strongest remaining evidence we have for the kind of meteorology Shaw may have conveyed in his lectures on meteorology, and later in the more general courses on thermodynamics he ran in the Cavendish. Beginning with observed phenomena, Shaw asked whether the motion of the clouds we see really indicates the motion of the air in which they are formed, and discussed examples ranging from a Cambridge fog to the development of a storm over the Gulf of Genoa. Then he introduced both tabletop experiments and diagrams of isothermals and isentropics to convey how physicists could contribute to meteorology. Shaw emphasized the simplifications and idealizations embodied in these tools and always integrated them into a discussion of clouds outside the laboratory. Delivering the lungs of the subject, he demonstrated, for example, both of the two main ways in which cloud could be formed by the mixing of currents of air of different temperatures, and by dynamical cooling. To illustrate the latter case, Shaw opened a flask of air to an evacuated chamber, showing that the size of the droplets formed depended on the number of nuclei present (which Shaw supplied by allowing the smoke of a match to pass into the globe).

Shaw commented that this subject—the study of the nuclei on which droplets of water formed—"almost belongs to Mr. John Aitken, who has founded upon the principles of it a method of counting the dust particles in the air."[21] But Shaw himself neglected Aitken's concern with counting dust particles, and focused instead on the interplay of temperature and pressure changes in rising air; the behavior of the drops formed on nuclei; and the possibility that different nuclei might come into play at different stages in the life-history of rising air. Significantly, Shaw suggested that after the first formation of cloud in rising air, further elevation would cause an increase in the size of the drops. Eventually this would diminish the power of the air to carry the drops (and nuclei) to the point where they would have to be left behind. Clouds would act like a filter: air rising above them would be free of nuclei and if condensation occurred a second time it must be under different conditions, "with a different set of nuclei."[22] Shaw commented:

> It would be interesting to know what depression of temperature is necessary to cause condensation upon nuclei of different characters, and what relation the temperatures at which condensation takes place bear to the ordinary dew point. These questions carry me, however, beyond my own knowledge and experience: I must limit myself to showing, later on, some experiments in connexion with the formation of coloured coronae that I think indicate a change in the character of the deposit with a change in the number or character of the nuclei.[23]

In contrast to Aitken's use of the cloud chamber to diagnose the state of air samples, Shaw's interest in cloud formation saw him sketch ways of moving between the details of laboratory research and natural processes. His illustrative and speculative discussion intermingled mimesis and manipulation, and posed questions about nuclei and cloud formation that he saw to be amenable to quantitative experimental study. By the time the article was published, Shaw could add a footnote substantiating this by referring to recent experiments from C.T.R. Wilson.[24]

C.T.R. WILSON "ON THE FORMATION OF CLOUD IN THE ABSENCE OF DUST"

Wilson came to Cambridge in 1888 and studied physics under Thomson, Glazebrook and Shaw. The professor taught Part I lectures (with Newall and Searle acting as demonstrators), and Glazebrook and Shaw joined Thomson in giving Part II lectures. Wilson later commented that Shaw "would always look on all sides of a question and sometimes did not commit himself to any conclusion."[25] After completing the Natural Sciences Tripos in 1892, Wilson spent two years demonstrating in the Part I laboratory and the Caius College chemical laboratory, and then taught briefly at Bradford Grammar School. His time with school children convinced him to come back to the Cavendish even without any security. Demonstrating physics for medical students gave him the association with the laboratory he required, while allowing him time to begin research on his own account. Wilson's laboratory notebook shows that in late March he began experiments on cloud formation.

Wilson never mentioned Shaw in this regard, but whether or not he heard Shaw in London, he is very likely to have seen Shaw's apparatus in the Cavendish and learnt his views. In any case, it is now clear that a tradition of meteorological physics existed within the Cavendish that gave a strong rationale for just the kind of experimental work that Wilson began a week after Shaw's lecture. That tradition surely provided fertile ground on which to foster the more distant—and more dramatic—inspiration of the mountaintop that Wilson explicitly referred to when describing the origins of his approach.

After a brief literature review, Wilson penned six questions on "Cloud Formation" that began with coronae and glories and finished by asking what expansion would be necessary to produce condensation in air free from nuclei.[26]

On the next page, headed "Experiments March 1895 CTRW," Wilson drew a diagram of an experimental arrangement that is as similar to Shaw's lecture cloud chamber as it is to the more elaborate apparatus with which Aitken counted dust particles. However, as Galison and Assmus observe, unlike earlier experimenters Wilson used a cotton wool filter to make sure that *all* the air going into the vessel could be filtered. (Shaw's arrangement also incorporated a filter, but at least when lecturing in London he actively supplied the globe with nuclei before carrying out a series of expansions.) For Galison and Assmus, this is the decisive point in Wilson's experimental procedure, an "artificial" aspect departing significantly from the mimetic tradition that initially inspired him. They ask why he would have used air that was "specially prepared for laboratory purposes" if his motivation was to reproduce natural clouds. Then they argue that the source of this "profound change in material culture and conceptual structure" was the analytic tradition of ionic physics practiced under the leadership of J.J. Thomson, and in particular Thomson's earlier theoretical study of the process of drop formation on ions.[27] But Shaw's theory of cloud formation gives a different motivation, which Wilson's notes confirm. Under the heading "Meteorological," two days later he sketched a theory of cloud formation like Shaw's and wrote that "Above the cloud the air will be supersaturated but devoid of nuclei."[28] Just how far the intimate details of empirical apparatus reach into natural processes clearly depends on the particular theoretical accounts in which they are embedded.

Soon Wilson had settled on a good arrangement. It featured a glass bulb inverted over a flask that was partially filled with water and sealed air-tight. Wilson did use specially prepared air, but he also deliberately avoided laboratory air, which (far from being ideal) might be contaminated by gases. Rather he made sure he had ordinary Cambridge air by leaning outside the window to collect his sample. Then, instead of passing it through a cotton wool filter, he carried out several preliminary expansions to remove all ordinary dust nuclei through condensation itself.[29] For Shaw and Wilson, this was undoubtedly a close analog to the action of clouds in the atmosphere. Neither could have regarded Wilson's experiments as universal in scope: his apparatus could only model one means of cloud formation, but that it allowed the reproduction of conditions that might be met in some circumstances in the ordinary atmosphere was clearly central to their understanding of the experiment.

As he proceeded, Wilson correlated the change in volume at which he observed the initial condensation, with the equivalent fall in temperature and change in vapor pressure, and calculated the size of the molecules involved. On April 3, he took the further step of working out how far dust-free supersaturated air would have to rise for condensation to take place. Wilson found that "At any height or pressure, if air is saturated at 16.2 [°C], it must rise 2.71 kilometres to produce condensation in absence of nuclei." Then he noted several meteorological implications:

> We might thus expect to find cloud layers separated by intervals of about this amount; the air between the cloud layers being supersaturated. This would only apply quantitatively at least, to the lower regions of atmosphere. In higher regions the temperature would be too low, probably, for water drops to be formed. The supersaturation required for ice crystals to be formed will probably be quite different. Direct experiment would be required to settle this as S[urface]T[ension] & c. of ice is quite unknown.
>
> It may be however that even at very low temperatures water drops and not ice are first formed. Above calculations would then apply even in the region of cirrus clouds, as subsequent freezing of drops could not affect supersaturation required for condensation.[30]

Wilson clearly regarded his experiments to move between the laboratory and nature. The validity of that move required both practical material steps (like gathering his air from just outside the laboratory window) and a network of more or less speculative theoretical considerations. With both practical and theoretical preconditions in place, Wilson could reliably bring the

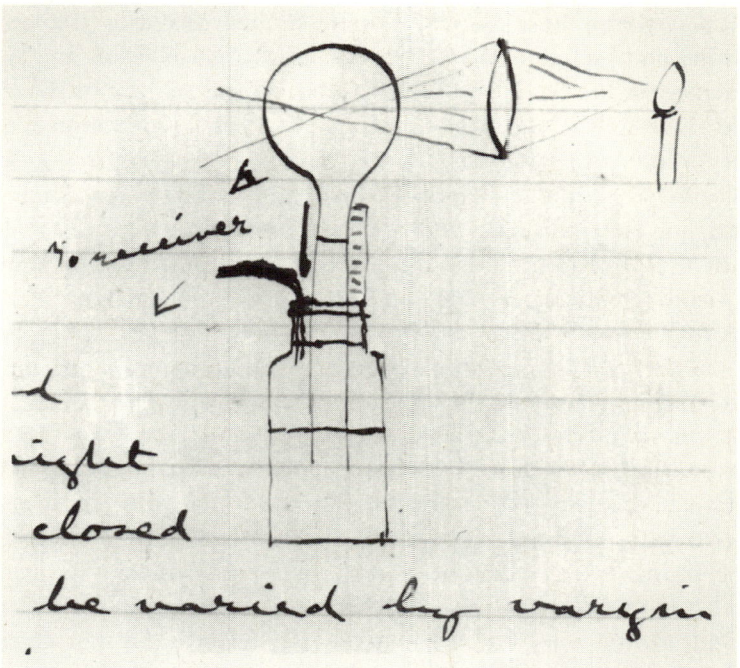

FIGURE 1   C. T. R. Wilson's notebook sketch of the apparatus he used for the first month of his experimentation on cloud formation (with some minor adaptations), from C. T. R. Wilson, Laboratory Notebook A1, 30 March 1895. © The Royal Society.

weather into his glass flasks, confident that his measurements correlated with natural phenomena. Almost a month after he had begun his experiments, Wilson took up his literature review again. Now he referred to J.J. Thomson's work for the first time and explicitly treated the condensation nuclei as ions: "If nuclei be present in shape of small electrified drops of radius 2×10^{-7}, each charged with atomic charge, we can calculate magnitude of this charge necessary to neutralize effect of S[urface]T[ension]."[31] Perhaps Thomson's account of ion drop formation had been the model for the condensation phenomena Wilson investigated from the very beginning—and Wilson was clearly keen to relate his experiments to his professor's matter physics. But it should now be clear that although he did not name Shaw as the source for his theory of cloud formation, meteorological and mimetic physics played a far more central role in the framing of Wilson's experiments and in his understanding of their implications than Galison and Assmus have suggested. For Shaw and Wilson, mimetic and analytic methods went hand in hand: Wilson's artificial clouds mimicked cloud formation in nature, but they could do far more than that alone.

Wilson gave a talk in Cambridge on May 13, and two years later published major papers in the *Philosophical Transactions* of the Royal Society.[32] By then he had used the cloud chamber to probe both the nature of the nuclei in his filtered air and of different gas samples. Several gases, but not hydrogen, gave a similar expansion ratio to that of air. Sunlight had no particular effect on condensation, but x-rays and uranium rays produced an increase of condensation at the same critical expansion. And there were two distinct phases of condensation. A light rain-like condensation appeared with expansion ratios above 1.25, but a much more abundant fog-like condensation began with ratios greater than 1.37. Wilson initially thought these two ratios might be the result of different constituents of the atmosphere carrying different quantities of charge, but following the suggestion of J.J. Thomson speculated that they were due to differences between negative and positive ions.[33] In a paper on "Rutherford and the Meteorologists," Arne Hessenbruch has shown that in many respects Wilson's cloud chamber work followed strategies that J.J. Thomson and Ernest Rutherford had recently developed in electrometer studies of gases exposed to both x-rays and uranium rays.[34] Exploring a further instance of research programs mixing meteorological and analytic resources, Hessenbruch indicates that a common concern with electrometer research and atmospheric electricity on the part of Elster and Geitel, Rutherford and C.T.R. Wilson between 1900 and 1901 led to Elster and Geitel's argument that leakage in electrometer devices indicated the presence of radioactive gases in the atmosphere. With the present chapter, Hessenbruch's study suggests a widespread recourse to meteorology in turn-of-the-century research programs that have commonly drawn attention largely for their implications for microphysics. Thus Wilson gradually extended the analytic reach of his device beyond the natural atmosphere—while leaving untouched the mimetic significance of his earliest work. His publications explicitly noted that the first expansion ratio would have a corollary in nature, but the second would not have any meteorological significance. Once drops had already begun to form, it was unlikely that any great degree of supersaturation could be maintained.[35]

MIMETIC METEOROLOGY AND ATMOSPHERIC ELECTRICITY: AITKEN'S CHALLENGE

Wilson soon developed a program of research in atmospheric electricity that further tightened the relations between mimesis and analysis in his work, and linked the Cavendish with field research and the electrograph records of Greenwich and Kew observatories. If rain brought negative ions to the earth—those Wilson's experiments indicated condensed with a lower expansion—that could explain the potential difference observed above the earth. He clearly regarded his experimental results to be fundamental to understanding the electrical effects of precipitation.[36] But their meteorological significance rested on the assumption that atmospheric condensation frequently takes place from a supersaturated condition. Serving as a referee for the Royal Society, John Aitken challenged Wilson on this. Aitken thought Wilson's research had "little or no bearing on meteorological phenomena," since its conditions were "never found in nature."[37] When the German physicists Elster and Geitel supported Wilson's views in 1900, Aitken fired off a letter to the journal *Nature* to state his opposition publicly. His views would make quite different features of both the natural environment on the one hand, and Wilson's apparatus on the other, pertinent to understanding whether laboratory manipulations correlated with natural processes.

Drawing on cloud observations and dust counter research, Aitken argued that only a small portion of the available dust particles ever become active centers of condensation in normal air; and that even if the air were to become supersaturated enough for ions to act as nuclei, the atmosphere would then be so unstable that cloud would never form. The extremely rapid expansions of Wilson's cloud chamber brought many nuclei into condensation at once. In contrast, in the atmosphere once any individual ion became active it would grow extremely rapidly, relieving the tension as it fell through the highly supersaturated air and soon falling to earth as a rain-drop.[38] For Aitken, Wilson's chamber couldn't approximate nature both because it used dust-free air and because the expansion was so rapid.

Wilson fought back. Using Aitken's own data, he cited cases in which very low counts of dust particles had been recorded in air close to the ground in places like Ben Nevis. Wilson thereby suggested that the mountain was quite like his laboratory flask (and that it was not only mountain haloes that were worth imitating), and argued that air at an even greater height was likely to be still less contaminated. In contrast to Aitken, he thought that a considerable vertical thickness of cloud was likely to remove any remaining dust particles by contact with the drops of water already formed. Aitken was not persuaded and sent another challenge suggesting that Wilson explore a number of issues concerning the presence and duration of ions in both pure and dusty air.[39] The exchange indicates that just where the boundary between the laboratory and nature was drawn would depend on the precise nature of the apparatus at issue, empirical measurements taken inside and outside the laboratory, and, even more importantly, on the network of assumptions within which those instruments and measurements were enmeshed.

Wilson continued to support his original views, referring to condensation from the supersaturated state in 1903 and continuing to focus on the limiting case of dust-free air in his electroscopic research, while Elster and Geitel carried out experiments on the conductivity of dusty air.[40] But eventually Napier Shaw came to accept Aitken's arguments. In his *Manual of Meteorology*, Shaw emphasized that the ordinary atmosphere contained an abundance of condensation nuclei, and that ions came into consideration only under specific laboratory conditions:

> The warning, to be repeated until it is generally understood, is that negative ions are not nearly such effective nuclei for condensation as the ordinary nuclei, whatever they may be, which are found in Aitken's dust-counter by rarefaction of the atmosphere.[41]

Thus the original view of cloud formation that inspired Wilson's research was discarded. As Britain's chief meteorologist, Shaw decided that in nature condensation never occurred from a supersaturated state.

Galison and Assmus have given a useful general framework by describing Wilson's work between 1895 and 1911 as "condensation physics" that straddled overlapping concerns in meteorology and ion physics. But they failed to notice the institutional basis for that mix in Cambridge itself. Perhaps this invisibility is a result of the longer-term institutional developments that saw meteorological research and other facets of cosmic physics increasingly removed from physics laboratories into other institutional homes. But it is also likely to be a result of Shaw's demeanor as a demonstrator and lecturer. Unlike the professor J.J. Thomson, Shaw trained students to ignore his own role. In this instance both personal modesty and the relative disciplinary profiles of physics and meteorology conspired to render both the meteorological dimensions of the Cavendish and Shaw's role in Wilson's research less visible than the distant mountain Wilson visited on vacations, or his laboratory professor.

In addition to the varied cultural priorities that lit up or shaded different aspects of the intimate mix of meteorology and ion physics in Wilson's work, I have shown that his research involved an extremely close and fruitful interplay between mimesis and analysis. Despite the substantial difference they recognized between the natural atmosphere and the atmosphere of the laboratory (whether that was ideal or contaminated), neither Shaw nor Wilson would have regarded mimesis and analysis as necessarily disparate endeavors. Even introducing cotton wool into his apparatus, Wilson believed he was mimicking nature. But he also pushed beyond the natural limit to analyze and dissect. The analytic reach of this research was based on Wilson's willingness to radically modify his apparatus. He used pure gases rather than common air, incorporated electric fields, and subjected larger samples to expansion; different protocols that might require months of work to reliably reproduce, and that on some occasions raised questions that remained unsettled for years. But on the whole, exploring material distinctions like this enabled Wilson to build arguments showing links between different phenomena. Showing that condensation occurred at similar expansion ratios helped argue for essential similarities in different constituents. Wilson's analysis generally served a synthetic understanding of nature, even across the divide between the laboratory and the natural atmosphere. The important point was to recognize in specific cases the particular work it would require to correlate the intimate details of laboratory manipulation with universal elements or natural processes.

NOTES

*For a more detailed version of this paper that explores instrumental issues more fully see Richard Staley, "Napier Shaw and the Invention of the Cloud Chamber," in *The Whipple Museum of the History of Science: Instruments and Interpretations*, ed. Liba Taub and Frances Willmoth (Cambridge: Whipple Museum with University of Cambridge Press, in press 2006).

1. J.M.C. Burton, "Pen Portraits of Presidents—Sir Napier Shaw, MA, ScD, LLD, FRS," *Weather* 50, no. 3 (1995): 89–93; J.M.C. Burton, "William Napier Shaw—Father of Modern Meteorology," *Weather* 59, no. 11 (2004): 307–308.

2. C.T.R. Wilson, "Reminiscences of My Early Years," *Notes and Records of the Royal Society of London* 14 (1960): 153–173, p. 166. See also C.T.R. Wilson, "Ben Nevis Sixty Years Ago," *Weather* 9 (1954): 309–311, p. 311; C.T.R.

Wilson, "On the Cloud Method of Making Visible Ions and the Tracks of Ionizing Particles (1927)," in *Nobel Lectures Including Presentation Speeches and Laureates' Biographies: Physics, 1922–1941* (Amsterdam: Elsevier, 1965), pp. 194–214, p. 194.

3. Peter Galison and Alexi Assmus, "Artificial Clouds, Real Particles," in *The Uses of Experiment*, ed. David Gooding, Trevor Pinch, and Simon Schaffer (Cambridge: Cambridge Univ. Press, 1989), pp. 225–274, on pp. 228–246 and pp. 246–249. A revised version is included in Peter Galison, *Image and Logic: The Material Culture of Microphysics* (Chicago: Univ. of Chicago Press, 1997), chap. 2.

4. The Biographical Press Agency, "William Napier Shaw, M.A., Sc.D., F.R.S." in Cambridge University Library Archives (CUL) ADD 8124 Box 2. On Shaw see also E. Gold, "Sir William Napier Shaw," *Quarterly Journal of the Royal Meteorological Society* 71 (1945): 192–194; E. Gold, "William Napier Shaw, 1854–1945," *Obituary Notices of Fellows of the Royal Society* 5 (1945): 203–230.

5. Shaw, "Twice Twenty and Four" typescript autobiography, "How I became a meteorologist," in CUL ADD 8124 Box 3, p. 8.

6. Shaw, "The Cambridge Attitude Towards Meteorology," CUL ADD 8434 Box 3.

7. Richard Tetley Glazebrook and William Napier Shaw, *Practical Physics*, Text-Books of Science Series (London: Longmans Green, 1885). See also Simon Schaffer, "Late Victorian Metrology and its Instrumentation: A Manufactory of Ohms," in *Invisible Connections: Instruments, Institutions, and Science*, ed. Robert Bud and Susan E. Cozzens (Bellingham, Wash.: SPIE Optical Engineering Press, 1992), pp. 23–56, pp. 37–38 and fig. 5.

8. Rayleigh to Shaw, 1 January 1885, CUL ADD 8434 Box 2. On Shaw and Glazebrook's role bridging different eras in the Cavendish see Dong-Won Kim, "J.J. Thomson and the Emergence of the Cavendish School, 1885–1900," *British Journal for the History of Science* 28 (1995): 191–226; Dong-Won Kim, *Leadership and Creativity: A History of the Cavendish Laboratory, 1871–1919*, vol. 5, *Archimedes* (Dordrecht/Boston: Kluwer, 2002) pp. 67–73.

9. Shaw's teaching is described by H.F. Newall in *A History of the Cavendish Laboratory* (London: Longmans, Green, and Co., 1910), pp. 110–113.

10. Ibid., p. 114.

11. W.N. Shaw, ed., *Practical Work at the Cavendish Laboratory: Heat* (Cambridge: Cambridge University Press, 1886).

12. Gold wrote that these men had vanished from the scene, and "could better have described [Shaw's] work and praised his virtues." Gold, "Sir William Napier Shaw," p. 192.

13. Robert E. Kohler, *Landscapes and Labscapes: Exploring the Lab-Field Border in Biology* (Chicago: University of Chicago Press, 2002), esp. pp. 11–19. See also

Henrika Kuklick and Robert E. Kohler, *Science in the Field*, vol. 11, *Osiris* (Chicago, Ill.: University of Chicago Press, 1996).

14. Galison discusses the cultural geography of condensation physics in Galison, *Image and Logic*, pp. 135–141.

15. For a broader study of the place of heroic study in Victorian culture see Bruce Hevly, "The Heroic Science of Glacier Motion," in *Science in the Field*, ed. Kuklick and Kohler, pp. 66–86.

16. Kohler, *Landscapes and Labscapes*, chap. 4.

17. The Meteorological Council of the Royal Society was instituted in 1878 and consisted of the Hydrographer of the Navy and five members nominated by the Royal Society. R.H. Scott served as secretary until Shaw succeeded him in 1900. See Gold, "William Napier Shaw, 1854–1945," pp. 208–209. Shaw's paper was W.N. Shaw, "Report on Hygrometric Methods; First Part, Including the Saturation Method and the Chemical Method, and Dew-point Instruments," *Philosophical Transactions A* 179 (1888): 73–149. For a discussion of different hygrometers see William Edgar Knowles Middleton, *Invention of the Meteorological Instruments* (Baltimore: Johns Hopkins Press, 1969), chap. 3.

18. Schaffer, "Late Victorian Metrology."

19. W.N. Shaw, "Report on the Present State of our Knowledge in Electrolysis and Electrochemistry," *Reports of the British Association for the Advancement of Science* 60 (1890): 185–223.

20. W.N. Shaw, "The Motion of Clouds Considered with Reference to their Mode of Formation," *Quarterly Journal of the Royal Meteorological Society* 21 (1895): 166–180, p. 167.

21. Ibid., p. 176.

22. Ibid., p. 178.

23. Ibid., p. 177.

24. Ibid. Since Shaw's lecture was published after Wilson had begun his experiments, it is possible he revised it in order to accommodate the implications of Wilson's experiments. I would argue, however, that Shaw's interest in the motion of air in cloud suggests that the line of thought originated with him. Note also that Shaw offers a fuller treatment of the subject than Wilson's notebook or his publications before 1900.

25. Wilson, "Reminiscences of My Early Years," p. 165.

26. Wilson reviewed papers by Jean Paul Coulier, Aitken and Robert von Helmholtz concerned with the propensity of air treated in different ways to form cloud in an expansion apparatus or steam jet, noting significant doubts that dust was required for droplet formation. His research notebooks are held at the Archives of the Royal Society, London, and have been indexed in P.I. Dee and T.W. Wormell, "An Index to C.T.R. Wilson's Laboratory Records and Notebooks in the Library of the Royal Society," *Notes and Records of the Royal Society of*

London 18 (1963): 54–66. His work on cloud chambers begins in Notebook A1 in a section headed "Cloud Formation &c." The literature review is followed by a list of questions that are designated as his own, in a section headed (on different sides of the page) "Cloud Formation &c." and "CTRW." His first dated experiments were made on March 26. In what follows, I shall refer to the notebooks by number: e.g. Wilson Notebook A1.

27. Galison and Assmus, "Artificial Clouds, Real Particles," pp. 244–246.
28. Wilson Notebook A1, March 27–28.
29. Aitken had established that this procedure removed dust. Later in his experimental work Wilson checked that it gave the same results as air filtered through cotton wool.
30. Wilson Notebook A1, April 3.
31. Wilson Notebook A1, April 22 (Wilson referred to Thomson by his initials as JJT). Galison and Assmus incorrectly state that this reference occurred within a week of the beginning of Wilson's experiments, Galison and Assmus, "Artificial Clouds, Real Particles," p. 249; Galison, *Image and Logic*, p. 99. Thomson treated ion drop formation in J.J. Thomson, "On the Effect of Electrification and Chemical Action on a Steam-Jet, and of Water-Vapour on the Discharge of Electricity Through Gases," *Philosophical Magazine* 36 (1893): 313–327.
32. C.T.R. Wilson, "On the Formation of Cloud in the Absence of Dust," *Proceedings of the Cambridge Philosophical Society* 8 (1895): 306; C.T.R. Wilson, "Condensation of Water Vapour in the Presence of Dust-Free Air and Other Gases," *Philosophical Transactions of the Royal Society of London A* 189 (1897): 265–307; C.T.R. Wilson, "On the Comparative Efficiency as Condensation Nuclei of Positively and Negatively Charged Ions," *Philosophical Transactions of the Royal Society of London A* 193 (1899): 289–308.
33. Wilson first thought different quantities of charge might explain the different ratios, but then noted Thomson's suggestion (and explored its meteorological application to explain the potential gradient observed between the atmosphere and the earth, described below). Wilson Notebook A8, 4 March 1898 and 7 July 1898. See Galison and Assmus, "Artificial Clouds, Real Particles," p. 251.
34. Thomson and Rutherford had focused on developing the electrometer as a detection device in radiation research. See Arne Hessenbruch, "Tempestas ex machina: Rutherford's Discharge Chamber and the Meteorologists, c. 1900," in *Radioactivity: History and Culture, 1896–1930s*, ed. John Krige and Christine Blondel (Cambridge, MA: Harvard Academic Publishers, forthcoming).
35. Wilson, "Condensation of Water Vapour," p. 286.
36. Soon after becoming a member of the Meteorological Council, Shaw brought Wilson's research to its attention. This resulted in funding for a project in which Wilson would compare Greenwich and Kew electrograph records, use balloons and electroscopes to take simultaneous measurements of the potential gradient at the earth's surface and well above it, and conduct experiments in the Caven-

dish to explore the nature of atmospheric electricity and its most likely sources. C.T.R. Wilson, "Preliminary Program of Work to be done on Atmospheric Electricity," *Report of the Meteorological Council to the Royal Society* (1899–1900): 110–111. Wilson referred explicitly to Thomson's suggestion concerning the source of different expansion ratios on p. 111.

37. Royal Society Archives R.R.14.273-4. Aitken was commenting on Wilson, "On the Comparative Efficiency as Condensation Nuclei of Positively and Negatively Charged Ions." His recommendation that the paper be moved to the less prestigious *Philosophical Proceedings* was not followed. Two years earlier, Aitken had enthusiastically supported the publication of Wilson's 1897 paper for introducing measurements into a field that had only been roughly sketched out previously: Royal Society Archives R.R.13.312-3.

38. John Aitken, "Atmospheric Electricity (A Letter to the Editor)," *Nature* 61 (1900): 514–515.

39. Ibid. 62: 366–368.

40. C.T.R. Wilson, "Atmospheric Electricity," Ibid. 68 (1903): 102–104.

41. W.N. Shaw, *Manual of Meteorology. Vol. 1: Meteorology in History*, 4 vols. (Cambridge: Cambridge University Press, 1926), p. 246.

CHAPTER 5

Scaling Down

The "Austrian" Climate between Empire and Republic

DEBORAH R. COEN

The *path of history* is in fact not that of a billiard ball,
which, once struck, follows a predictable course, but
resembles rather the path of a cloud, which also follows
the laws of physics but is equally influenced by something
that can only be called a coinciding of facts. For
everywhere the wind blows from e[ast] to w[est], because
in the e[ast] there is a ma[ximum] and in the w[est] a
mi[nimum] in air pressure; but that a place lies between
both, where no nearby mountain mass diverts the air's
course, or where otherwise competing influences make
themselves felt, all these circumstances constituting the
weather: all these elements, when they come together,
even if they are calculable, are really facts and not laws.

—Robert Musil, 1923[1]

The history of meteorology in Central Europe seems at first glance to
be an exemplary instance of the modern progression from local
to global science: from a geographically limited study of local phenomena to an international project to understand atmospheric phenomena
on a global scale. Weather science in Central Europe boasts all the familiar
trappings of such a history: a centralizing bureaucracy, a vast empire, an

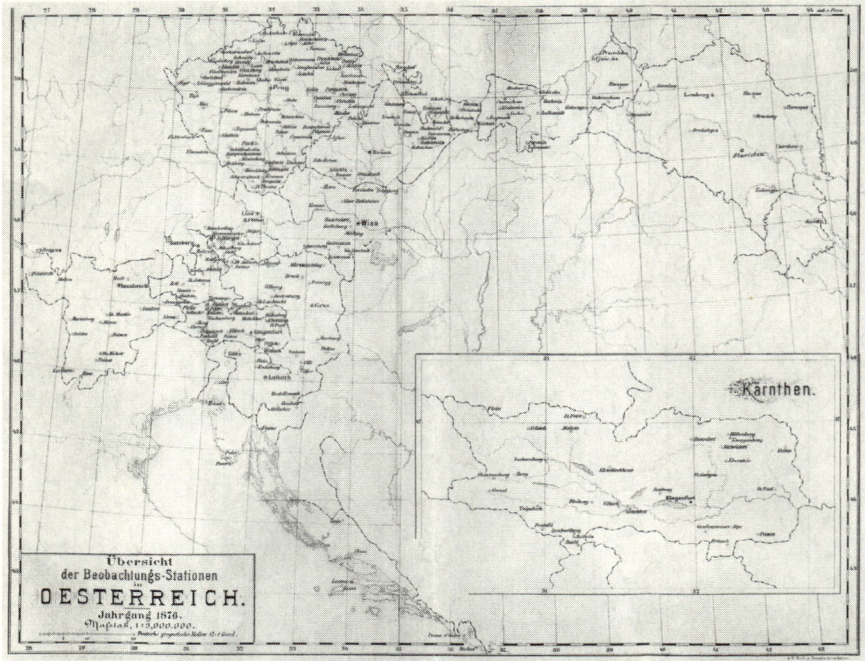

FIGURE 1   The distribution of meteorological observing stations in the Austro-Hungarian Empire, 1876. Source: Christa Hammerl, et al., eds., *Die Zentralanstalt für Meteorologie und Geodynamik, 1851–2001* (Graz: Leykam, 2001), p. 57. Courtesy of the Zentralanstalt für Meteorologie und Geodynamik, Vienna.

expanding telegraphic network, and sophisticated statistical and carto-graphic methods. Vienna's Central Institute for Meteorology and Geophysics, founded in 1851, coordinated observations transmitted telegraphically from stations sprouting throughout the Hapsburg Empire. During the second half of the nineteenth century, the institute expanded its network of observing stations across the empire's vast territory, from the Alps to the steppes of present-day Ukraine. Meanwhile, meteorologists in Vienna developed new ways to put this influx of data to use: in 1865 the Central Institute began to produce a daily weather map, and in 1877 it began transmitting daily fore-casts for the empire. In 1884, the Austrian Meteorological Society began publishing the *Meteorologische Zeitschrift*, the most prestigious journal in the field throughout the first half of the twentieth century. By the turn of the century, the Viennese were renowned experts in the methods of large-scale or *synoptic* mapping for the purposes of forecasting and theoretical analysis.

The trouble with meteorology's familiar narrative of globalization is that it omits crucial innovations of the past while blinding us to potentials for the future. Indeed, in recent years, the Swiss-based group CLEAR has argued that the key to meeting the challenge of global climate change lies in *regional* strategies of research and reform. These scientists point to the Alps as the region that best exemplifies the benefits of small-scale climate analysis. And they emphasize the value of alternating between local and global perspectives or "up-and-down-scaling."[2] Prescient as their program may be for the policies of the twenty-first century, it is merely an echo of the lessons of Austrian climatology in the 1920s.

The Austria that fostered this flexible science was not a mighty empire, but rather the rump state remaining after the treaty of St. Germain in 1919. Vienna's Central Institute lost the better part of the old imperial observing network. Moreover, the old crown lands now beyond Austria's borders demanded the return of their weather records. Austrian meteorology had lost almost all its advantages of scale. If globalization were really the driving force in the history of modern weather science, then meteorology in an amputated Austria should have floundered. Intriguingly, it flourished. Austrian meteorologists succeeded because they learned to shift strategically between global and local perspectives.[3]

At first, in the last months of the Great War, the outlook for these scientists was bleak. The new Austria was cut off from the agricultural center and sources of raw materials of the former empire, and it was responsible for war indemnities. Inflation and food shortages struck the Viennese hard. The director of the central institute, Felix Exner, was obliged to ask the state for money just to keep the institute's employees from starving. In 1920, the institute begged the American Meteorological Society for assistance. Aid trickled in.[4]

Beyond these material obstacles, the new Austrian nation was in the throes of an identity crisis. As historians have recently argued, a rump state like the First Austrian Republic (1918–1934) faces challenges unlike anything that befell post-colonial Britain or France.[5] George Clemenceau expressed this memorably at St. Germain when he referred to the new Austria as "what was left over." Austrians themselves wanted no part of the new republic; many hoped to merge with Germany. Well after peace was signed, border disputes continued in the Sudetenland, in Carinthia, and in South

FIGURE 2   The Dissolution of the Habsburg Empire after World War I. Source: Barbara Jelavich, *Modern Austria: Empire and Republic, 1815–1986* (Cambridge: Cambridge University Press, 1987), p. 157. Copyright © 1987 Cambridge University Press. Reprinted with the permission of Cambridge University Press.

Tyrol. Austria in the 1920s was radically split between its progressive metropolis and its Catholic, tradition-bound alpine hinterland. Vienna became Europe's model of a successful socialist experiment, with thriving programs of public education, health care, and housing.[6] But the power of the socialist capital was precariously balanced against that of clerical-conservatives and right-wing paramilitary organizations in the provinces.[7] Accentuating the opposition between worker and peasant was the comparatively late commercialization of Austrian agriculture.[8] The divide between "Red Vienna" and the conservative provinces posed a unique obstacle to the work of nation-building. Unlike England and France, where national movements were primarily urban, and unlike Eastern Europe, where nationalism took root in peasant culture, interwar Austria "oscillated" between its urban and rural identities.[9] Meanwhile, many Austrians still saw their borders as crudely artificial.

So it was with some irony that, during and after the Great War, Vienna's Central Institute requested state support to continue a pre-war series of monographs on "The Climatography of Austria." It was in no way clear what this title now meant. Consider for instance the task facing Austrian geophysicist Victor Conrad, who was responsible during the war for a volume on "The Climatography of Bukowina," a region lying in present-day Ukraine. Conrad tried to adhere closely to the analytical model set by the Austrian pioneer of meteorological statistics Julius Hann in the first volume of the series. In order to reduce measurements from different stations to the same time period, Hann had advised referring to a local "normal" station, the central node of a regional network within the larger imperial system. But the war made it impossible for Conrad to access large portions of the local records. Not even for one station in the region did he have sufficiently complete data. As his predicament suggests, the statistical methods of climatology in Austria in 1917 were the product of a hierarchical imperial administration.[10]

In pre-1918 Austria, climatology had been an imperial science. Had Austrian meteorologists relied exclusively on methods developed before the war, they would have gotten nowhere in the early 1920s. Instead, the Central Institute's main interwar directors Felix Exner (1916–1930) and Wilhelm Schmidt (1930–1936) led their colleagues in new directions. Soon after the war's end, Exner made sure that the new government recognized the peacetime value of a science that had been vital to the Austrian war effort. In a letter of 1919, Exner addressed himself simultaneously to the ministries of transportation, defense, justice, trade, agriculture, and health. He heralded the value of his institute for air travel, agriculture and forestry, hydrography, mining, wind power, legal decisions in tort cases, hygiene, transportation and tourism.[11]

Still, *proving* this potential would require fundamental shifts in the practice of meteorology at the Central Institute. No longer could Vienna's meteorologists employ the methods that had served an empire. In the terms of the years to come, they would supplement "*Großraum*" (large-scale) meteorology with research at the micro- and meso-scales, and they would address new questions. In what sense were local climates independent of one another, they asked themselves. Where did the urban climate end and the mountain climate begin? Faced with the problem of analyzing a new entity, the "Austrian climate," meteorologists responded by defining a gradation of scales of analysis, from the intimate to the global.

THE ORIGINS OF A NEW
CLIMATOLOGY I: MOUNTAINS

The meteorologists of the First Republic chose a research strategy that sacrificed geographic breadth in favor of detail. They sought to cover the Alps with a thick net of observing stations, channeling their resources into "mountain meteorology." Their program might be called "post-imperial" for its use of intensive rather than extensive methods. As Director Exner repeatedly and unabashedly reminded the authorities, research on alpine weather was a good investment. Physicians had recognized climate as a factor in the treatment of consumption since the nineteenth century, although they made little effort to distinguish the effects of climate from those of diet, exercise, or even the very act of traveling. There was as yet no agreement on which climate was most beneficial, nor had there been much effort to study the physical characteristics of therapeutic climates.[12] Then, in the first decade of the twentieth century, the Swiss physician Carl Dorno popularized the treatment of tuberculosis patients with sunlight at Davos, the site of Thomas Mann's *Magic Mountain*.[13] When Austria was hit with post-war inflation, its tuberculosis patients could no longer afford a Swiss resort. The Austrian Society for the Control of Tuberculosis hunted for a domestic alternative. For Exner's new program in mountain meteorology, the ultimate pay-off would be the discovery of an "Austrian Davos": an exceptionally sunny locale, at least by Central European standards.[14] Immediately after the peace was signed, Exner and his colleague Heinrich Ficker began a campaign to locate a suitable spot for a sun-therapy center within Austria's new borders. With this justification, they erected new alpine observing stations "of the first order." In the *Meteorologische Zeitschrift* in 1921, Ficker insisted that the eastern Alps contained enough sunny peaks to treat "the suffering people from all over Europe." It was "positively the duty of the experts to point to this expansive area, the suitability of which for the purposes of high-altitude therapy is unmatched by any other region in the remaining German Alps."[15] Note Ficker's reference to the "German Alps." On one hand, "German" stood opposed to Italian-speaking, since one candidate for such a sanitorium lay in South Tyrol, which Austria had been forced to cede to Italy. At a more basic level, though, Ficker's choice of phrase shows that Austria's scientists in 1921 could not assume an Austrian identity apart from Germany. The national climate was as yet ill-defined.

Over the next few years, Austrian scientists broadened their concept of "climate therapy." By 1928, meteorologists Viktor Conrad and Walter

Hausmann had developed an extensive list of requirements for certifying a locale as a "*Luftkurort*," a climatically therapeutic site. The criteria ranged from the intensity of sunlight to temperature, humidity, rain- and snowfall, to the color of the sky.[16] The program was carried out under the aegis of the Public Health Office of the Federal Ministry of Social Administration. Often the process began when a community appealed to the Central Institute for consideration. A scientist would then travel to inspect the area. If his initial assessment was favorable, he would order the installation of a "climatic observation station," consisting ideally of eight instruments, to be operated by local volunteers.[17] So, for instance, the town of Mönichkirchen, with moderate summer and winter temperatures, average cloud cover, but heavy snowfall, learned in 1929 that it had been approved as a destination for convalescents, but not for the seriously ill—not, for instance, for tuberculosis patients.[18] For more fortunate locales, the meteorological stations furnished material that could be used directly for advertising. In the words of the unfortunately named director of marketing for Austrian tourism Dr. Erwin Naswetter ("Dr. Wetweather"), "One can expect from it a scientific support for the entire marketing program, taking into account above all the modern view of advertising that sees the objectivity and truth of the commendation as the chief requirement of a good piece of publicity."[19] As others in the tourist industry now stressed, the marketing of climates would now rely on scientific facts: "Speculation must be excluded."[20]

The new climate therapy influenced the local climates to which ailing Europeans chose to expose themselves, but it also changed *how* they experienced those climates. As Gregg Mitman observes, personal accounts by "health-seeking" tourists are a window onto developing environmental sensibilities and perceptions of place.[21] Self-observation became a keystone of the new medical climatology. In the late '20s, for instance, Viktor Conrad introduced the *Befindenskarte* or "health map." According to his plan, physicians in various locations would choose "suitable" patients to record their own state of health three times a day. The average results of each locale would then be plotted on a weather map, producing over time a visual image of climatic healing and deterioration. This was an undertaking in the tradition of what Nicolaas Rupke terms "Humboldtean medicine," which visualized geographical distributions of diseases and correlated them with climate.[22] Conrad foresaw that it would be possible to read off these maps the effect of various climatic elements, such as polar and equatorial air, on the health of patients.[23] What distinguished Conrad's project from other contemporary initiatives to map bioclimatological data was his emphasis on self-observation. Appealing to physicians, he suggested that the maps be used to track the "subjective" change in the health of patients in the course of their cures. Intriguingly, he expected that patients' errors in reporting their states would be random. Yet one might wonder whether the new self-consciousness of the patients' experience of weather might not alter its effects. Certainly the experience of weather was becoming ever more disciplined. Consider an article that appeared in 1937 in the *Bioklimatische Beiblätter*, the scholarly journal founded in 1934 as a spin-off of the *Meteorologische Zeitschrift* by Schmidt and Linke. The author, identified only by his initials, was particularly "weather sensitive," according to the terminology of the day. He presented the meticulous observations he had made since 1919 of the dependence of his health on the weather. As he put it, "self observation" had become "habit."[24]

The medical-climatological initiative in Austria became the model for other such programs in Europe in the 1930s. And the Austrian government was celebrated internationally for its support of medical climatology. German climatologist Friedrich Linke complained in the late 1920s of the Weimar government's indifference to health-related climatological research. Linke pointed to the Austrians and Czechoslovaks as the pioneers of this valuable new field. Although on a smaller scale than in Austria, the Czechoslovaks had maintained the Habsburg tradition of climatological research. Yet they too had reoriented the discipline around local needs, specifically agriculture and the expansion of medical spas. In 1928, the Balneological Congress held in Baden officially thanked the Austrian and Czechoslovakian governments for their support of medical climatology, or "bioclimatology," as it was increasingly known.[25]

The status of bioclimatology in the two former Habsburg states is revealing. While the relationship of science to overseas imperialism is now a well-studied subject, little research has been done on the role that scientific institutions play in the process of post-imperial nation-building in former continental empires (those of the Habsburgs, Romanovs, or Ottomans). Interwar Austria and Czechoslovakia, formerly metropole and province, had each become a post-imperial state adjusting to borders widely perceived as arbitrary. For these nations, sciences of the *local* seem to have held special potential. Local science offered short-term practical pay-offs. As we will see, it also took on the more ambiguous task of describing the new nation to itself.[26]

Austrian weather scientists and their allies in the tourist industry intended the measurements from the new alpine stations to lure the healthy as well as the sick. Evidence of a locale's suitability for summer and winter sports could be just as lucrative as a designation as a medical spa.[27] Tourism played a vital if fraught role in the interwar Austrian economy. During the hyperinflation of the early '20s, Austrians did not hide their resentment of the foreigners who flocked to the mountains and lakes of the Tyrol and Salzkammergut to take advantage of the falling crown.[28] Yet it was clear that in the new Austria, tourism would be a matter of survival. The "golden age" of the Austrian tourist industry stretched from the stabilization of the Austrian economy in 1923 to the stock market crash of '29.[29] New alpine resorts opened year-round to take advantage of the growing popularity of skiing as an adventure sport (slopes and lifts did not yet exist).[30] The late '20s likewise saw the beginning of a boom in the construction of mountain roads, opening up many of Austria's most famous peaks to automobiles. Most were designed as scenic routes, showcases of Austria's natural beauty. Even more dramatically, despite the economic crisis of the '30s, twelve painstakingly built cable cars opened to the public between 1926 and 1937.[31]

Riding this trend, in 1924 the Central Institute began issuing reports on snow conditions at the major ski areas, and in 1930 it instituted the "Alpine Weather Service," providing forecasts for the mountains by radio.[32] At first these appeared to be valuable incentives to winter travel. Before long, however, a conflict emerged between the aims of the Meteorological Institute and those of the tourist industry. In one instance of an enduring dispute, the Institute's forecasts were blamed in 1934 for a slow-down of summer tourism in the Salzkammergut, the small region of lakes and hills outside Salzburg. This region's Union of Spas and Summer Resorts protested that the Institute had repeatedly singled out the Salzkammergut in its forecasts of bad weather, while rarely mentioning the region when good weather was in store. The union also complained that the meteorological institute too often erred in its forecasts of rain or clouds on Sundays. "That these forecasts of rain are not conducive to tourism undoubtedly need not be emphasized."[33] In its defense, the Meteorological Institute provided the Ministry of Education with a multi-year statistical analysis of forecasts, weather, and tourism. These showed that the actual weekend weather, not the forecasts, was to blame for the empty rooms in the Salzkammergut. The Institute assured the ministry, moreover, that its forecasts had erred on the side of optimism, ever since its one glaring mistake—a pessimistic forecast for Easter, 1930 (which still had not slowed the tourist traffic that year).[34] Director Schmidt explained to the Ministry that the Central Institute felt deeply its duty to the nation, but it remained "a scientific institution." "With all due consideration of the grave economic crisis from which our homeland suffers, we must nonetheless refuse to publish intentionally false forecasts . . ."[35] Ironically, the institute was implicitly arguing that the public paid little attention to its forecasts. After insisting so firmly on the practical value of their work in the 1920s, Austria's meteorologists now hesitated to be judged according to such utilitarian standards. In this small, post-imperial nation, the scientists' challenge was to prove their value to the state without wholly sacrificing their independence.

THE ORIGINS OF A NEW CLIMATOLOGY II: CITY

In addition to boosting tourism and promoting medical spas in the Alps, Vienna's climatologists found ways to aid the new state closer to home. In the immediate aftermath of the Great War, tuberculosis haunted working-class neighborhoods in the cities. The Ministry of Sanitation and Health blamed the "wild settlements" that had sprouted in the post-war years for fostering disease. These communities lacked proper drainage, sewage, and sunlight. As one report to the Ministry attested, "Poor living and family conditions are fertile soil for the physical and moral degeneration of youth."[36] Among a population for whom a mountain cure was out of reach, might climatology still prove useful?

The 1920s in Red Vienna was the heyday of modernist projects to redesign the city on a rational basis. Thus the intended audience for the new urban climatology included hygienists, architects, and engineers. As Schmidt and Ernst Brezina wrote in their 1937 survey of the subject, "Its task is to correctly influence municipal housing politics and to create more bearable living conditions for the urban population, [conditions] which correspond better to the physical and spiritual nature of man than is usually the case for the urban dweller today."[37] Previously in Austria-Hungary, climatological data on urban centers had been collected within the imperial framework. The measurements of the urban observatory were compared with those at outlying observatories, yielding no more than a rough characterization of the features specific to the urban climate. After the war, just as the climatologists of the First Republic replaced imperial with local methods in the Alps, they proposed a new and emphatically local scheme for urban climatology. As Viennese climatologist Friedrich Steinhauser reflected in 1934, "Investigations of urban temperature patterns gained practical significance once value came to be placed on hygenic housing design and the functional design of communities. What was then needed was to invent new methods of investigation."[38]

One such method was initiated by Wilhelm Schmidt in 1927 with funding from the *Notgemeinschaft der Deutschen Wissenschaften* and the ministries of agriculture and education. Schmidt's "mobile climatological observatory" was actually no more than a large Opel fitted out with meteorological instruments. But it allowed Schmidt and his colleagues to take measurements closely spaced in time at locations throughout the city and its outskirts. Schmidt was the first meteorologist to use such a "moving laboratory."[39] The lessons of small-scale research in the Alps could now be put to use in the metropolis: "The question was now attacked from an entirely different perspective, as the results of the detailed investigations of microclimatology turned widespread attention to the characteristics of the climate of different types of locales, of small-scale land formations, of varied conditions of vegetation . . ."[40] Just as Austria's climatologists had become sensitive to the challenges of studying the manifold microclimates within a mountain range, they recognized that enclosed urban spaces required new methods of research. "What remains of air movement within enclosed spaces—mostly in the form of drafts or light breezes—is of an essentially different magnitude as what we observe outdoors. The usual methods of meteorology fail here; those that replace them are indirect or highly sensitive."[41] Indoor climates had drawn scientific and medical attention since the eighteenth century for their influences on human health, as Vladimir Jankovic shows in his contribution to this volume. With the introduction of commercial air-conditioning at the turn of the twentieth century, engineers further developed the study of what they called "artificial climates," yet their methods relied heavily on incommunicable personal experience.[42] The Austrians sought instead to develop an empirical physical science of "artificial climates." These ranged from a city to a neighborhood to a single building or room, down to the micro-scale of the space between a person's clothing and his bare skin. A researcher might even track his own "personal climate" during his daily routine as he made his way from home to streetcar to laboratory and back.[43] Tellingly, one technique that did carry over from mountain to urban climatology was the Beaufort scale of wind strengths, the lower end of which could capture the slightest draft in a sealed room. At the same time, less familiar meteorological variables took on new significance in the urban environment, such as levels of carbon dioxide, dust, radiation, and atmospheric electricity.

Like their eighteenth-century English counterparts, the Austrians focused on the urban climate as it was experienced by the city's inhabitants, particularly in its impact on health. Schmidt and Brezina were convinced that Vienna's climate was partly responsible for the difficulty of raising healthy children in the city, and thus for the weakening of the urban population over time. They came to the conclusion that pre-war efforts to ease the effects of life in an "artificial climate" through seasonal adjustments were fruitless. The solution they endorsed was more radical: to provide homes for urban workers just beyond the city itself. Schmidt and Brezina thus aligned their research with the "settlement" or "green city" movements sprouting up in Austria and Germany at this time. They expected that the newly built communities would effect a transformation in the urban populace at once physical, psychological, and political: "to strip the inhabitants of the city of their extreme urban character, to make of urban dwellers people who no longer think and feel exclusively urbanly, for whom nature is no longer an object for occasional rejuvenation on Sundays, but who are instead tied to nature like their forefathers who worked the land." By growing their own food on their own small patch of land, workers would benefit "physically" and "spiritually," and they would be better able to withstand economic crises like that of the early '30s. Such advantages could only come from the natural environment, the authors stressed, not from a seasonally adjusted artificial climate. Finally, the new suburban communities promised to "bring contentment to part of the population in which bitterness has dominated, in some cases for generations; to give them once again a positive attitude towards their own nation and people."[44] With this conclusion, the authors made clear that urban climatology was a contribution to the politics of nation-building.

THE NATIONAL CLIMATE

But a contribution of what kind? The urban climatologists demonstrated that Austria's climate—in the political as well as physical sense—was shaped by its cities, not just its mountains. Working-class life, their studies implied, was as constitutive of Austrian identity as *Lederhosen* and *Dirndl*. These studies implicitly opposed the image of Austria being promoted on the political right—that of an alpine nation of tradition-bound villages.[45] Ironically, however, these same scientists had been instrumental in constructing that quite different picture of their nation. Conservative Austrian statesmen of the '30s praised in *völkisch* terms the very successes that climatologists had helped make possible, from new scenic routes to international tourist destinations in the Alps.[46] Thanks in part to the climatologists' strenuous promotion of the Austrian Alps, the quaint alpine village now stood for all the world as the symbol of the new nation. Moreover, urban climatologists echoed the rhetoric of the political right when they prescribed the emulation of rural life as a cure for a diseased urban mentality. Even as they sought to protect the interests of urban workers, they seemed to promote what modernist architect Adolf Loos called "Verdorfung"—the attempt to give urban centers the feel of villages that now existed only as nostalgic ideals.[47]

In fact, this ambiguity was symptomatic of left-wing politics in the republic. Historians have stressed the paradoxes that distinguished Austro-Marxism, from its conflicted relationship to the state to the paternalism that clouded the relationship of the leaders to the workers. Anson Rabinbach has even argued that the very successes of Austrian socialists in the areas of education and culture paved the way for fatal discord within the party itself. Thus, rather than a tug-of-war between two competing versions of national identity, one urban and one alpine, we should not be surprised to find in public discourse in the First Republic an ambivalent amalgam of the two.[48]

The complexity of the discourse on climate emerged, for instance, in the effort to establish a summer drama festival in Salzburg shortly after the war. The festival was to project a specifically Austro-German, Catholic, and conservative national identity even as it announced the post-war renewal of international cultural life. The festival's promoters, including Hugo von Hofmannsthal, enlisted nationalism and cosmopolitanism as unlikely allies. They portrayed Salzburg as Europe's natural (and now cultural) *center*, the heart of *Mitteleuropa*. Yet, as Hofmannsthal described in a promotional essay, Salzburg's landscape conveyed a specifically Austro-German character: "A valley is a natural theater, and remarkably, the theatrical drive of the south-German race follows the mountain chains . . . There is nothing accidental, it is a geographical reality, a profound connection between what seems wholly spiritual and what seems wholly physical . . ."[49] Hofmannsthal celebrated in particular the "natural" effect that this site had on its foreign visitors. In his deft prose, he portrayed the Salzburg landscape as the embodiment of both seclusion and centrality, province and metropolis, the national and the cosmopolitan.

Cultural programs supported by the state government likewise promoted an *alpine* national identity. They sought to bring the country to the city dweller and the city dweller to the country. Government-sponsored courses taught urban workers about their alpine homeland, while the *Heimatschutz* conservation movement pressed citizens to use their free time

127

to experience the "authentic" Austria first-hand. Hiking in fact became "the single most preferred noncommercial form of recreation" among Vienna's adult workers.[50] Yet the mere act of hiking did not carry an overt political message; the pasttime meshed just as well with socialist culture as it did with right-wing nationalism.[51] The political lessons of the landscape had to be made explicit. What then was Austrian about the "Austrian Alps"?

Here, Vienna's climatologists helped furnish an answer. "The Austrian Alps" was the title of a Volksbildung course given at the University of Vienna in 1926–7. The presentations spanned the Alps' geology and vegetation, its music, costumes, and art. As a whole, the lectures conveyed to an urban audience the message that Austria's culture and mountain landscape were intimately linked. Felix Exner's contribution on "The Climate of the Alps" made this point directly. Exner asked his audience to imagine how the climate of Central Europe would look if the Alps ran not west-east but north-south, like the Rocky mountains. On the east side, Austria's side, the west wind would come down dry and produce a dusty plain like that east of the Rockies. "So we may thank God that the Alps run from west to east; otherwise Central Europe would not be a fertile region which for millenia has sustained us and allowed us to advance."[52] Exner himself had been exploring the complex interaction between the climate and geomorphology of the Alps in experiments at the Meteorological Institute. By carving "river valleys" a millimeter wide and heaping sand along them, he had managed to model several of the Austrian Alps' most famous peaks. In "The Climate of the Alps," however, he was asking not just how earth and wind had shaped each other, but how, in Hofmannsthal's terms, the physical and spiritual had co-evolved.

Like Hofmannsthal, Exner used the Alps in the early '20s to stage an elaborate representation of a new Austrian identity, one that brought nationalism and cosmopolitanism into uneasy alliance. In 1922, he strategically arranged for Austria to host one of the first post-war pan-European meteorology conferences. It would be held at the legendary Sonnblick observatory, 3100 meters high in the Alps. The observatory was a feat of engineering and fund-raising in the 1880s, and a symbol of the ambitions of Austrian meteorology in the nineteenth century. Now, in the wake of the World War, Austria was bankrupt and disgraced. Exner, Ficker and their colleagues were desperate to rescue their standing in the international scientific community. But the meeting would be in October and the mountain already blanketed in snow. The steep ascent and several nights' rough lodging would be a challenge even in good weather. Ficker explained that the conference would be a declaration to the rest of Europe: "Just let someone come and say that Austria lacks initiative and guts."[53] When Ficker set out from Graz, it was pouring rain, and the conference participants began their hike under heavy clouds. As Ficker joked, on most hikes the presence of a meteorologist would be considered bad luck—and here there were 25 of them. But fortune smiled on them. The sun shone, and the conference was a roaring success.

But it was also an event that displayed in high contrast the contradictions of science in post-1919 Austria. The conference was to provide an opportunity for discussion of the latest theories of weather on vast scales from continents to hemispheres. Yet the Austrians had picked a mountain range, a region in which weather could not be explained without detailed knowledge of the local geomorphology. Exner and his Austrian colleagues virtually defied their guests to ask for a weather forecast before setting out. Locals, they implied, still knew best.

A SCIENCE OF THE LOCAL

Much as climatologists contributed at a practical and ideological level to the First Republic, the political situation proved surprisingly conducive to scientific innovation. The pragmatic decision to study the Austrian climate within its shrunken borders launched a new science, which the Austrians christened *Kleinklimatologie*. They were not the first to study local climates; scientists in Russia and the United States had been doing so for years. The botanist Gregor Kraus had laid a groundwork for *Kleinklimatologie* in 1911 with his observations of a startling diversity of micro-climates in air lying no more than two meters above the shell-limestone formations near Würzburg.[54] Yet it was the Austrians who gave the field of local climatology its theoretical and methodological grounding, while teasing out its relevance to research on a global scale.

As Wilhelm Schmidt portrayed it in the 1930s, *Kleinklimatologie* was the *natural* outgrowth of the dramatic shift in the nation's contours. The new Austria, with its predominantly mountainous landscape, demanded a new approach to climate study. "The image of the climate thus receives in detail its peculiar imprint, it no longer fits into the familiar large scale, it demands methods of observation, analysis, and representation different from those we are used to."[55] As Schmidt explained, the Austrian Republic was ideally suited to small-scale climatology for three reasons. First, its mountainous terrain made the spatial distribution of weather phenomena particularly apparent, since weather changes rapidly in the vertical direction. In Austria, weather "plays out spatially [*räumlich abspielt*]." These spatial variations were so pronounced in Austria, Schmidt posited, that it was hard to overlook their effects on vegetation and even on the conditions of one's own life. Moreover, mountains create their own climates. Thus, a second reason to make small-scale climatology an Austrian specialty was the presence of unique phenomena, such as the Föhn, the seasonal alpine wind. In the end, though, research on the small scale had become a necessity. Since Austria was now located almost entirely within the Alps, its scientists could not obtain an overview of their effects. They could study the alpine climate *only* in its partial, local effects.

The new field inverted the familiar strategies of imperial science at every step. To the Austrian practitioners of the new *Kleinklimatologie*, large-scale methods now looked "clumsy." Their research instead prioritized the local, the fluctuating, and the heterogeneous, all the qualities that imperial meteorologists had pared away to create the smooth and constant *overview* of weather on a continental scale. The meteorologists of imperial Vienna had valued geographic breadth, spacing their observatories 40 or more kilometers apart. The meteorologists of the Republic would instead measure a few "typical" locales as densely as possible, making the best of their shrunken borders. For this purpose, permanent observatories were impractical, and Schmidt's "mobile observatory" proved crucial. Not only the distance of measurements from each other but also their distance from the ground was critical. Large-scale climate science had always circumvented the variations caused by local topography by measuring temperatures two meters above ground level (and wind measurements at six meters). Research on the small scale, on the other hand, would study the air closest to the ground (*die bodennahe Luftschicht*), the atmosphere that humans actually breathed. Where *Großraum* research had relied on long-term mean values to cancel out fluctuations of the meteorological elements, small-scale science was interested in short-term variability, in the natural extremes that shaped the lives of plants, animals, and humans alike. After all, one early frost was meaningless in the long term but potentially life-threatening in the short. Finally, where the imperial observatories had been limited to measuring a few basic elements, small-scale measurements would encompass new factors crucial to health and agriculture, such as dust, ultraviolet radiation, and carbon dioxide.

Schmidt and his colleagues sought to delimit the new field of *local* or *mesoclimatology (Kleinklimatologie)* from the sciences of both the macro and the micro. Yet their many attempts to define these scales proved unstable.[56]

Large-scale climatology, they argued, had grown out of man's experience of travelling the world and its goal was the "study of the climate of a *land-scape*." A *Kleinklima*, by contrast, covered only *part* of a landscape, a single valley or cliff, for instance. Even smaller, a *Mikroklima* could not even be recognized as part of a landscape; it was, to first approximation, merely flat. Yet it turned out that just as there could be a local climatology of a cave or of a city, there could equally be a micro-climatology of such spaces. One way of understanding the difference between these was methodological. The instruments of standard, large-scale climatology could be applied to the meso-scale, but as one approached the micro-scale they would begin to produce "a noticeable disturbance in the air under investigation."[57] Elsewhere, however, Schmidt insisted that new instruments were equally essential for the study of the meso-scale as for microclimatology. Indeed, the sensitivity of climate on the smaller scales to human activities provided a further means of differentiation; thus, the macro-scale was naturally set apart from the micro- and meso- because the former was "natural," the latter "artificial." Such a definition suggested that phenomena at one scale were to some extent isolated from those at another. In fact, as we will see below, these same scientists defined "dependent" and "independent" climates. Ultimately, Schmidt himself acknowledged the instability of such definitions. The point was not to make dogmatic distinctions but to provide guidelines for measurement. Indeed, by demonstrating the difficulties of delimiting one scale from another, the Austrians productively highlighted the interrelations among local, regional, and global phenomena.

By no means did Austrian meteorology turn its back on the *global* science of weather. As Schmidt argued to the Ministry of Education on his institute's behalf, the same science that served local needs also fostered international ties:

> Of all Austria's scientific establishments the Central Institute embraces not merely the widest circle of beneficiaries, from the simplest farmer and tourists to researchers studying the conditions of life; it also certainly has the broadest and strongest international connections: our weather and climate are of course part of the general circulation of the whole earth . . .[58]

One might think that these were no more than the exaggerated claims of a scientist desperate for funding. Yet their validity is beyond doubt: Austrian meteorologists succeeded in making their local studies of global relevance.

A prime example is Schmidt's own work on the turbulent exchange (*Austausch*) of energy between air masses (as well as in water). Helmholtz had described atmospheric turbulence qualitatively in the 1880s, but attempts to quantify such phenomena were not made until circa World War I by British scientists G.I. Taylor and Lewis Richardson.[59] Ludwig Prandtl had derived as early as 1904 equations describing the formation of vortices in mixing layers, but these were little known before the 1920s.[60] Schmidt's approach instead benefited from the work of his Austrian colleagues Exner and Margules, who had begun before the war to study the effects of discontinuities between air masses—well before Vilhelm Bjerknes conceived his theory of the "polar front."[61]

Schmidt's research carried the physics of *Austausch* from the smallest scales to the largest. On one hand, this was an eminently practical endeavor. He addressed his findings to meteorologists, oceanographers, geographers, physicists, botanists, and farmers. For the most part, he dealt with *Austausch* phenomena perpendicular to the ground, through which warmer air layers higher in the atmosphere mixed with cooler layers lower down. Such phenomena, he made clear, provided "important foundations for the characterization of the 'local climate.' "[62] For instance, valleys were marked by a smaller degree of *Austausch* (expressed as a lower coefficient A) than mountain peaks and ridges. Valleys, therefore, experienced greater disparities between high and low temperatures both diurnally and annually. Likewise, this explained why the climate of a forest clearing (low *Austausch*) could be considered "independent," while that of an open field (high *Austausch*) was "dependent."

But *Austausch* phenomena did not only affect climate on the small scale. Schmidt's definition of the *Austausch* coefficient was equally applicable to *horizontal* energy transport exchange in the atmosphere, i.e. to the transport of air across the earth's surface. Treating global effects with tools developed for the local scale, Schmidt was able to visualize how climatic fluctuations could travel from equatorial regions towards the poles, losing intensity as they went.[63] In this way, Schmidt demonstrated the significance of his "local" climatology for global weather forecasting, as well as for fundamental theoretical issues at the vanguard of atmospheric physics and hydrodynamics in the 1920s. Indeed, Schmidt's theory of *Austausch* was quickly absorbed into the international canon of dynamical meteorology.[64]

The local climatology pioneered in Austria became far more than a local science. Throughout Central Europe in the 1930s, disciplines including medicine, agriculture, anthropology, and geography incorporated the lessons of small-scale climate science. In the assessment of one popular German survey of weather science from 1937: "*Kleinklimatologie*, the theory invented by meteorologists of climate on the smallest scale, has already placed in the hands of many neighboring sciences the most valuable material for clarifying many open questions. For plants, animals, and people it has introduced us to the climate of life itself."[65] In the Third Reich, Willy Hellpach appropriated the new climate research to support his claims for the influence of climate on culture. Through the many editions of his book *Geopsyche*, the notion of local climatic influences on individuals and communities reached a popular audience.[66] Another German who recognized the import of *Kleinklimatologie* was Albrecht Haushofer, who served as a foreign policy consultant to the Nazi government before turning against Hitler in a failed asassination attempt of 1941.[67] Haushofer's theoretical and diplomatic work were both concerned with the problem of *Lebensraum* or "living space," a Darwinian concept developed by the geographer Friedrich Ratzel and adopted by National Socialists to justify Germany's predatory designs on Eastern Europe. For the analysis of a local climate as *Lebensraum*, Haushofer insisted, conventional averages were meaningless compared to knowledge of the "fundamental dynamics" of the atmosphere and of a wide variety of meteorological factors.[68] These, of course, were precisely the methodological lessons of climate study on the small scale. According to the Munich meteorologist Rudolf Geiger, writing in 1941, the very existence of the new science of microclimatology supported the ideology of Nazi imperialism. It was "no accident," he contended, "that microclimatology has been developed in Germany," for the quest to understand and exploit the local climate had been driven by "the scarcity of living space [*Lebensraum*]."[69] At this point, local climatology had come full circle: fostered by a post-imperial process of down-scaling, it had come to serve expansionist ambitions of unprecedented ruthlessness.

CONCLUSION

It appears that the question of the European: What am I? really means: Where am I? It is not a matter of a phase in a process governed by laws, and not a matter of destiny, but simply of a situation.

—*Robert Musil, 1923*[70]

133

Historians and political scientists have judged institutional continuity to be a key factor in the post-imperial politics of the former centers of contiguous empires. Karen Barkey argues that the survival of imperial institutions in states like interwar Austria allowed elites to guide "ideology and politics" and made an "assimilationist" (as opposed to nationalist) politics more likely.[71] The case of Vienna's Central Institute for Meteorology and Geophysics suggests that the role of imperial institutions in this transition has not always been so simple. As the institute adapted to the new state, discontinuities in its mission were as striking as continuities. Moreover, as its twin programs of mountain and urban climatology suggest, the ideological implications of its research were highly malleable, as likely to be nationalist as cosmopolitan, and just as useful to socialists as to the former elites.

In the science of *Kleinklimatologie*, Austrians developed a resource for investigating, imagining, and transforming space that was uniquely suited to a *Kleinstaat*, a small nation. They pioneered a science with the potential to guide the exploitation of what little land remained, while giving "natural" meaning to seemingly arbitrary borders. In just five years, for instance, they helped instill meaning (and fortune) into the notion of the "Austrian Alps." Yet this science of the local was hardly provincial. It traveled well beyond Austria's borders, winning new respect for a dishonored nation. More profoundly, it provoked reflections on the very meaning of the local. By providing a framework for analyzing atmospheric phenomena on vastly different scales, the new science fostered a new spatiality. It gave experts and laypeople alike new ways to think about their experience of weather as multilayered, composed of some elements that were strictly personal and others that were local, regional, or global. To what extent were these elements independent? How did they interact? These were questions that propelled basic research in climatology in this period, but they also exercised the popular imagination.

Robert Musil, for instance, enlisted the science of weather to visualize the situation of the "German" in the post-World War I world. How did the individual fit into history's course? Musil envisioned man as a particle of air caught between wind currents, his path hinging on the infinite contingencies of pressure, temperature, and topography. The particle's identity was equivalent to its trajectory, past and future, and its trajectory in turn was a function of its instantaneous position in space. One was not "destined" but "situated"; place replaced fate. Two years earlier, recalling the advent of war in 1914, Musil had likewise turned to atmospheric metaphors: "One suddenly became a tiny particle humbly dissolved in a suprapersonal event and, enclosed by the nation, sensed the nation in an absolutely physical way." He described the nation as an "enormous heterogeneous mass . . . that oscillates between solid and fluid," stressing the "atmospherically undefined nature" of the experience of patriotism.[72] These images vividly conveyed his skepticism towards the concepts of "nation" and "race" in the wake of the Great War. Yet, if the nation was "a fantasy," Musil's task was to interpret the fantasy's power, to identify the images it evoked. For Musil, images of atmospheric processes furnished a physical sense of the individual's relationship to the imagined space of the nation and to historical time.

Musil was hardly alone in expressing spatiality and temporality in atmospheric terms. In another exemplary novel of the First Republic, the socialist writer Rudolf Brunngraber used the image of a night wind to describe a psychological transformation in the wake of the Great War. Brunngraber's hero develops an urge to travel widely, a trait the author deems symptomatic of the age: "The longing to be far away [*der Trieb ins Weite*] seemed to have become an epidemic in the '20s."[73] This *Wanderlust* grows in a series of moments when Karl becomes acutely aware of the scale of the world. In each case, he likens the experience to the sensation of wind: "Then he suddenly felt the size of the world, as if he stood outside in the night wind." He feels a mere current of air in the motionless metropolis as a "wind of existence." Like Musil, Brunngraber suggested that atmospheric phenomena had the capacity to awaken a new geographic and historical consciousness in Central Europe.

At the other end of the political spectrum was Hofmannsthal, the dramatist of the Austrian Alps. Known for his personal sensitivity to weather, Hofmannsthal used terms like "atmosphere," "air," and "circulation" to describe what he famously called the "spiritual space of the nation." For Hofmannsthal, whose prose gave the alpine landscape the voice of the Austrian *Volk*, the language of weather performed the ultimate literary feat: it decisively collapsed the opposition between the physical (*physikalisch*) and the spiritual (*geistlich*) character of the nation.[74]

In fact, in German-speaking Central Europe between the world wars, space itself was being imagined anew. The new technologies, theories, and visions of space in interwar Central Europe have yet to be surveyed by historians. In a region that had been remapped wholesale by the peace treaties, scholars began to question the very meaning of *Raum* (space). Disciplines from philosophy to geography, biology, and history contributed to this discussion. Even "the daily press and popular magazines are filled with '*Raum*' as a catchword," wrote a German geographer in 1932.[75] In the Third Reich,

judges one historian, "spatial concepts belonged to the everday political vocabulary."[76] Nazi sympathizers and others spoke of a "Volk ohne Raum,"
of "Lebensraum," "Blut und Boden," and of the "Bodenpolitik der Zukunft."
In this context, personal, quotidian space was as disputed as the space of
geopolitics. The political left and right seized on utopian urban planning as
a means of transforming society by remaking the space of everyday life.
Even the space of childhood was reconceived, as the child's objective *Lebensraum* (*der Raum, in dem das Kind lebt*) was distinguished from his subjective experience of space (*der Raum, den das Kind erlebt*).[77] These new visions
of space in turn proved decisive for the theory and practice of *Weltpolitik*
(geopolitics) under Hitler. These varied participants were simultaneously re-
defining the space that constituted "Central Europe" and "space" as Central
Europeans experienced it. They were also grappling with the problem at the
core of the geographical enterprise in interwar Central Europe, that of rec-
onciling scientific-material and idealist-cultural modes of explanation.[78]
What climatology contributed to this process was an infinitely adaptable
means of imagining the relationship between culture and physical landscape.

NOTES

The author thanks Christa Hammerl, Vladimir Jankovic, Conevery Bolton Valenčius,
and Nasser Zakariya for helpful conversations and comments on this essay.

Abbreviations: AdR = Archiv der Republik, AVA = Allgemeine Verwaltungsarchiv,
Österreichisches Staatsarchiv, Vienna, ZAMG = Zentralanstalt für Meteorologie
und Geodynamik.

1. Robert Musil, "The German as Symptom," in *Precision and Soul*, ed. David S.
   Luft and Burton Pike (Chicago: University of Chicago Press, 1990) 150–192, on
   169.
2. Peter Cebon et al., eds., *Views from the Alps: Regional Perspectives on Climate
   Change* (Cambridge, Mass.: MIT Press, 1998). Sarah Strauss, "Weather Wise:
   Speaking Folklore to Science in Leukerbad," in Ben Orlove et al., eds., *Weather,
   Climate, Culture* (Oxford: Berg, 2003), 39–60.
3. For Felix Exner's efforts to prevent this, see AVA Unterricht 4A (ZAMG,
   1918–1927), 1918: document 1277, p. 7.
4. Christa Hammerl, "Die Geschichte der Zentralanstalt für Meteorologie und
   Geodynamik," in C. Hammerl, W. Lenhardt, et al., eds., *Die Zentralanstalt für
   Meteorologie und Geodynamik 1851–2001* (Graz: Leykam, 2001), 19–299, on
   133–135.
5. Karen Barkey, "Thinking About Consequences of Empire," and E. J. Hogsbawm,

"The End of Empires," in *After Empire: Multi-Ethnic Societies and Nation-Building* (Boulder, Colo.: Westview Press, 1997), 99–114, 12–18.

6. On Austro-Marxism's successes and failures, see in particular Anson Rabinbach, *The Crisis of Austrian Socialism* (Chicago: University of Chicago Press, 1983).

7. Martin Kitchen, *The Coming of Austrian Fascism* (Montreal: McGill-Queen's University Press, 1980).

8. Herbert Nikitsch, introduction to *Schönes Österreich: Heimatschutz Zwischen Ästhetik und Ideologie* (Wien: Österreichisches Museum fur Volkskunde, 1995), 13–29, on 24.

9. Reinhard Johler, "Das österreichische: Vom Schönen in Natur, Volk, und Geschichte," in *Schönes Österreich*, 31–41, on 34–5.

10. For Felix Exner's petition to continue publication of the *Klimatographie* series see AVA Unterricht 4A, 1918: document 1277, and 1920: document 929. Viktor Conrad, *Die Klimatographie der Bukowina* (Vienna: Gevold & Co., 1917).

11. AVA Unterricht 4A, document 267, 29 March 1920.

12. Sheila M. Rothman, *Living in the Shadow of Death: Tuberculosis and the Social Experience of Illness in America* (Baltimore: Johns Hopkins University Press, 1995), 21, 195, 200.

13. See the obituary for Dorno in *Bioklimatische Beiblätter* 9 (1942): 1–3. For the social history of tuberculosis sanitoria in Germany see Flurin Condrau, *Lungenheilanstalt und Patientenschcksal* (Göttingen: Vandenhoeck & Ruprecht, 2000).

14. Felix Exner, "Das erste Beobachtungsjahr auf der steiermärkischen Sonnenheilstätte Stolzalpe (1200 m)," *Meteorologische Zeitschrift* 39 (1922), 149–152.

15. Heinrich Ficker, "Wo findet man in der deutsch-österreichischen Alpen einen Ersatz für Davos?" *Meteorologische Zeitschrift* 38 (1921), 307–309, on 309.

16. Conrad and Hausmann, "Meteorologische Gesichtspunkte bei der Beurteilung alpiner klimatisch-therapeutischer Ortslagen," in *Mitteilungen des Volksgesundheitsamtes* (1928), 1–16.

17. Ibid. and reports of observations in AdR Soz. Volksgesundheit: Kl. Beob. St.

18. V. Conrad to R. Höfler, AdR Kurorte 1929, document 32976.

19. AdR Kl Beob St 1930, document 51584.

20. Report of meeting of study commission for Hochserfaus, 7 Jan. 1927, AdR Kurorte 1927, document 21913.

21. Gregg Mitman, "Health, Leisure, and Place in Gilded-Age America," *Bulletin of the History of Medicine* (2003): 600–635.

22. Nicolaas A. Rupke, "Humboldtean Medicine," *Medical History* 40 (1996): 293–310.

23. Conrad, "Bericht über die bisherigen Ergebnisse der medizinisch-klimatischen Aktion des Volksgesundheitsamtes im Bundesministerium für soziale Verwaltung," *Mitteilungen des Volksgesundheitsamtes* (1929): 3–11, on 7. See also Conrad to the Volksgesundheitsamt, 3 July 1929, AdR Kl. Beob. St. 1929, document 2491.

24. N.N., "Jahreszeitlich bedingte Änderungen physiologischer Verhältnisse einer besonders wetterempfindlichen Person," *Bioklimatische Beiblätter* 4 (1937): 106–8.

25. Report from forty-third Balneological Congress Baden, AdR Kurorte 1928, document 17591. Walter Hausmann, "Grundlagen und Organisation der lichtklimatischen Forschung in ihrer Beziehung zur öffentlichen Gesundheitspflege," *Mitteilungen des Volksgesundheitsamtes* (1932): 1–20; and reports from the 1932 meeting of the radiation committee of the International Meteorological Organization, AdR Kl. Beob. St. 1933.

26. On the significance of localized climate science for the development of regional identity in the pre-Civil War American South, see Conevery Bolton Valenčius, *The Health of the Country: How American Settlers Understood Themselves and Their Land* (New York: Basic Books, 2002), chapter 4.

27. AdR Soz. Volksgesundheit, Kurorte 1927: document 21913.

28. Robert Kriechbaumer, ed., *Der Geschmack der Vergänglichkeit: Jüdische Sommerfrische in Salzburg* (Vienna: Böhlau, 2002).

29. Roman Sandgruber, *Ökonomie und Politik: Österreichs Wirtschaftsgeschichte vom Mittelalter bis zur Gegenwart* (Vienna: Ueberreuter, 1995), 376.

30. In a one-year period from 1920 to 1921 the membership of the Austrian Ski Association grew from 4.3 to 7.7 million. Jürgen Schwab, *Soziokulturelle und ökonomische Aspekte der Entwicklung des alpinen Skisports in Österreich* (Linz: Universitätsverlag Rudolf Trauner, 1999), 45, 51.

31. Georg Rigele, "Sommeralpen-Winteralpen," in *Umweltgeschichte: Zum historischen Verhältnis von Gesellschaft und Natur* (Vienna: ÖBV, 2000), 121–145.

32. Hammerl, "Geschichte der ZAMG," 142.

33. AVA Unterricht 4A: ZAMG, document 17247-32 (14 June 1932).

34. AVA Unterricht 4A: ZAMG, document 17247-32 (14 June 1932).

35. AVA Unterricht 4A: ZAMG, document 27289-34 (7 Sept. 1934).

36. AdR Soz. Sanitäts und Gesundheitswesen, 1935, document 46839.

37. Brezina and Schmidt, *Das künstliche Klima in der Umgebung des Menschen* (Stuttgart: Enke, 1937), 207.

38. Steinhauser, "Großstadttrübung und Strahlungsklima," *Bioklimatische Beiblätter* 3 (1934): 105–111, on 105.

39. Rudolf Geiger, *The Climate Near the Ground*, trans. M. N. Stewart (Cambridge, Mass.: Harvard Univ. Press, 1950), 379.

40. Steinhauser, "Großstadttrübung und Strahlungsklima," *Bioklimatische Beiblätter* 3 (1934): 105–111, on 105.

41. Brezina and Schmidt, *Das künstliche Klima*, 3.

42. Gail Cooper, "Custom Design, Engineering Guarantees, and Unpatentable Data: The Air Conditioning Industry, 1902–1935," *Technology and Culture* 35 (1994): 506–536.

43. Brezina and Schmidt, *Das künstliche Klima*, 130.
44. Brezina and Schmidt, *Das künstliche Klima*, 207.
45. Georg Rigele, "Sommeralpen-Winteralpen," in *Umweltgeschichte: Zum historischen Verhältnis von Gesellschaft und Natur* (Vienna: ÖBV, 2000), 121–145.
46. Ibid. and Kriechbaumer, ed., *Geschmack.*
47. Reinhard Johler, "Das Österreichische: Vom Schönen in Natur, Volk, und Geschichte," in *Schönes Österreich*, 31–41.
48. Rabinbach, *Crisis of Austrian Socialism*; Helmut Gruber, *Red Vienna* (New York: Oxford University Press, 1991).
49. Hofmannsthal, "Festspiele in Salzburg," in *Gesammelte Werke* (Frankfurt: Fischer Taschenbuch, 1979) 9, 269–271, or 266.
50. Gruber, *Red Vienna*, 121.
51. Dagmar Günther, *Wandern und Sozialismus: Zur Geschichte des Touristenvereins "Die Naturfreunde" im Kaiserreich und in der Weimarer Republik* (Hamburg: Verlag Dr. Kovac, 2003).
52. Felix Exner, "Klima der Alpen," in *Die österreichische Alpen* (Vienna: Franz Deuticke 1927), 165–175, on 167.
53. Reinhard Böhm, *Der Sonnblick* (Vienna: Östereichischer Bundesverlag, 1986), 56, cited in Hammerl, "Geschichte der ZAMG," 139.
54. See for instance Charles F. Brooks, "How may one define and study local climates?" in *Comptes Rendus du Congrès International de Géographie* (Paris: Librairie Armand Colin, 1931), Vol. 2, 1–10. Gregor Kraus, *Boden und Klima auf kleinsteim Raum* (Jena: Fischer, 1911). Theodor Homén in Finland had studied the temperatures above different types of earth and published *Der tägliche Wärmeumsatz im Boden und der Wärmestrahl zwischen Himmel und Erde* (Leipzig, 1897). Kraus' and Homén's results had been discussed in the principal German-language meteorology textbook of the day, Julius Hann's *Lehrbuch der Meteorologie* (Leipzig: Tauchnitz, 2nd edition 1906), 37.
55. Schmidt, "Kleinklimatische Beobachtungen in Österreich" *Geographischer Jahresbericht aus Österreich* 16 (1933): 42–72, on 43.
56. Geiger and Schmidt, "Einheitliche Bezeichnungen in kleinklimatischer und mikroklimatischer Forschung," *Bioklimatische Beiblätter* 4 (1934): 153–156.
57. Ibid., 153.
58. Schmidt, 23 June 1934, AVA Unterricht 4A 1934, document 20375.
59. Frederik Nebeker, *Calculating the Weather: Meteorology in the Twentieth Century* (San Diego: Academic Press, 1995), 33.
60. Roddam Narasimha, "Divide, Conquer and Unify," *Nature* 432 (2004): 807.
61. H. C. Davies, "Vienna and the Founding of Dynamical Meteorology," in Hammerl, ed., *Zentralanstalt*, 301–312.
62. Wilhelm Schmidt, *Der Massenaustausch in freier Luft und verwandte Erscheinungen* (Hamburg: Henri Grand, 1925), 26.
63. Schmidt, *Massenaustausch*, 109.

64. David Brunt, *Physical and Dynamical Meteorology* (Cambridge: Cambridge University Press, 1939), 231.

65. Johannes Grunow, *Wetter und Klima: Ihr Wirken und ihre Beziehungen zur lebenden Welt* (Berlin: Wegweiser, 1937), 216.

66. A contemporaneous work of popularization was the multi-authored *Klima, Wetter, Mensch*, ed. H. Woltereck (Leipzig: Quelle & Meyer, 1938), to which the Austrian Brezina contributed.

67. Albrecht was the son of Karl Haushofer, the founder of the German school of geopolitics, who was a mentor to Rudolf Hess, one of Hitler's early close associates in the Nazi Party.

68. A. Haushofer, *Allgemeine Politische Geographie und Geopolitik*, vol. 1 (Heidelberg: K. Vowinckel, 1951), 85–87. The treatise was published a decade after Haushofer's execution.

69. Rudolf Geiger, *Das Klima der bodennahen Luftschicht: ein Lehrbuch der Mikroklimatologie*, second edition (Braunschweig: Vieweg & Sohn, 1942), 4.

70. See note 1.

71. Karen Barkey, "Thinking About Consequences of Empire," in Mark van Hagen and Karen Barkey, eds., *After Empire: Multiethnic Societies and Nation-Building* (Boudler, Colo.: Westview Press, 1997), 99–114, on 106–7.

72. Musil, "The 'Nation' as Ideal and as Reality" (1921), in *Precision and Soul*, 101–116, on 103 and 111.

73. Brunngraber, *Karl und das 20. Jahrhundert* (Göttingen: Steidl, 1999), 162.

74. Karlheinz Rossbacher, *Literatur und Bürgertum: Fünf Wiener jüdische Familien von der liberalen Ära zum Fin de Siecle* (Wien: Böhlau, 2003), 467; Hofmannsthal, "Festspiele in Salzburg," and "Das Schriftum als geistiger Raum der Nation," in *Gesammelte Werke* 9, 227–245, esp. 235–36.

75. Albrecht Haushofer, "Zur Problematik des Raumbegriffs," *Zeitschrift für Geopolitik* 9 (1932): 723–734, on 723.

76. Mechtild Rössler, *Wissenschaft und Lebensraum: Geographische Ostforschung im Nationalsozialismus* (Berlin: D. Reimer, 1990), 2.

77. Martha Muchow, *Lebensraum des Großstadtkindes* (Hamburg: M. Riegel, 1935).

78. David Livingstone, *The Geographical Tradition* (Cambridge, Mass.: Blackwell Publishers, 1993), 177ff; Woodruff D. Smith, *Ideological Origins of Nazi Imperialism* (New York: Oxford University Press, 1986), 91 and Chapter 7. For examples of geographers grappling with this problem in this period see Hugo Hassinger, *Geographische Grundlagen der Geschichte* (Freisburg: Herder, 1953), esp. 6; Willy Hellpach, *Geopsyche* (Leipzig: W. Engelmann, 1939), esp. 213, 246; Giselher Wirsing, *Zwischeneuropa und die deutsche Zukunft* (Jena: E. Diederichs, 1932), 9–10.

CHAPTER 6

Teaching the Weather Cadet Generation

Aviation, Pedagogy and Aspirations to a Universal Meteorology in America, 1920–1950

ROGER TURNER*

Clouds of carbon dioxide ice crystals scudded across the sky. At another time, the cyclone blowing at 55° N 10° E might have been a weak nor'easter. But for the readers of "Some Aspects of the Meteorology of Mars," in the February, 1950 issue of the *Journal of Meteorology*, storm fronts sweeping across Mars were a logical, if surprising, product of three decades of pedagogical and institutional development.[1] Beginning in the 1920s, American meteorology students learned that a universal ideal lay at the heart of their emerging profession. Meteorology was the science of atmospheres. By 1950, this community could recognize that weather events on Mars, like weather on Earth, should be understood as particular cases of the general laws that governed all atmospheres.

The discoverer of that Martian nor'easter was Seymour Hess. While his studies of Mars were "generally considered 'far out' figuratively as well as literally" by his colleagues, he was representative of the new American meteorological community. He was one of roughly 6,000 "weather cadets," young men trained to forecast weather for the United States Army Air Force and Navy during World War II. Picked out by the training program's leader,

141

Carl-Gustaf Rossby, as one of the brightest in his class, Lt. Hess spent the war at the University of Chicago. He taught subsequent classes of cadets how to calculate the movement of air masses, locate frontal systems, and predict conditions in the upper atmosphere. Like hundreds of his peers, he completed an advanced degree after the war, then spent a successful career developing the "young science" of meteorology. Hess helped establish a department of meteorology at Florida State University and later worked extensively with NASA.[2] His peers founded dozens of other academic departments, forecast weather for the military around the globe, managed weather conditions for airlines, integrated radar, satellites and computer modeling into the daily work of the Weather Bureau, and generally made secure, middle-class lives from the science they had first learned during the war. As Figure 1 shows, the weather cadet generation dominated the demography of American meteorology for decades.

To understand this post-war community of scientists, we need to recognize that it was actively constructed by an earlier generation of meteorologists. The leaders of this earlier generation, Carl-Gustaf Rossby and Francis Reichelderfer, are now celebrated as the founders of modern American meteorology.[3] Between the 1920s and the 1940s, Rossby, Reichelderfer and their allies designed the institutions, established the curriculum, and cultivated the values that guided the weather cadets trained during World War II. Understanding their agenda within the social and political context of the interwar years reveals why the weather cadet generation was taught to aspire to produce a universal science of the atmosphere.

Most accounts of this period focus on the introduction and acceptance of the "Bergen School" into American meteorology.[4] They explore how the Norwegian concepts of air masses, fronts, and the genesis of cyclones were slowly incorporated into American meteorology. The central puzzle in several accounts is why these concepts, now so familiar to meteorologists and the television-watching public, were resisted by the US Weather Bureau well into the 1930s—despite the charm and boyish enthusiasm of Bergenite Carl-Gustaf Rossby, on his way to becoming the foremost theoretical meteorologist of his era. Memoirs written by meteorologists who were young during the 1930s and 1940s solve this puzzle by blaming the senior forecasters who ran the Weather Bureau. Poorly educated bureaucrats who stubbornly clung to obsolete methods, these old men did not understand the math and physics behind the Bergen School, and were blind to the path-breaking research that would transform weather forecasting from an art to a science.[5]

142

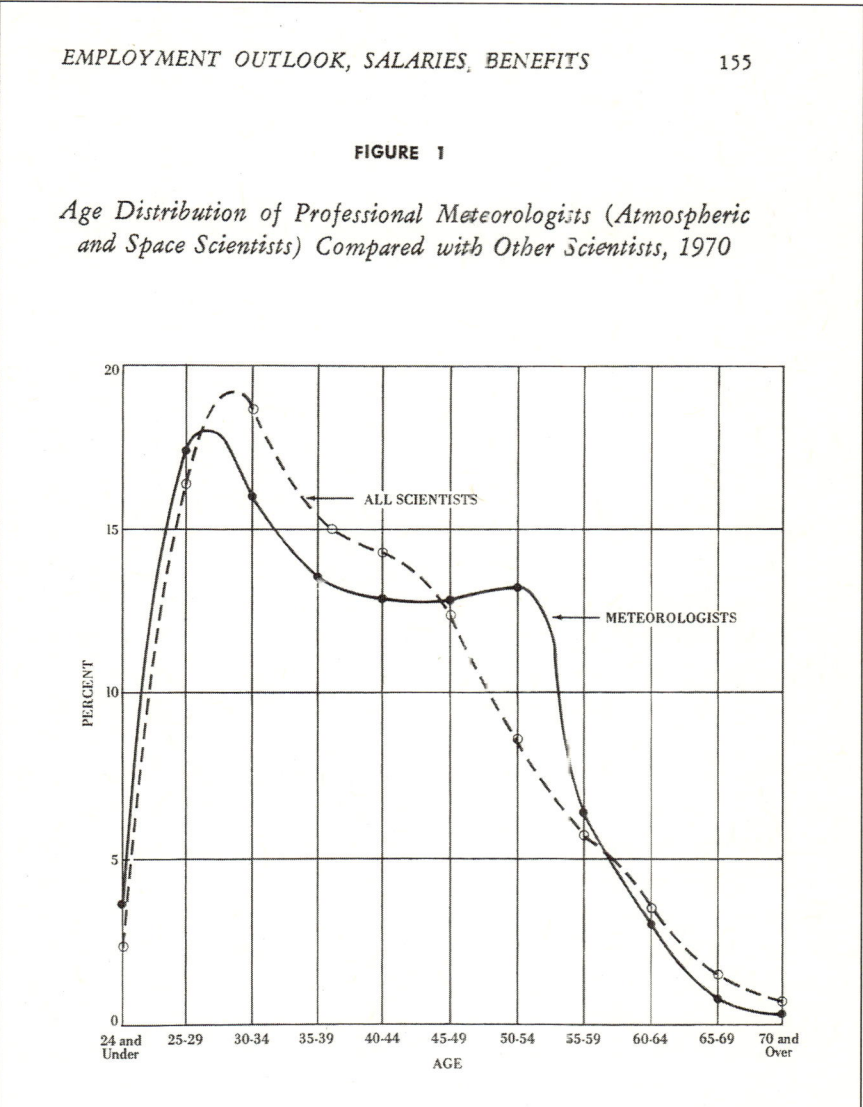

FIGURE 1

Age Distribution of Professional Meteorologists (Atmospheric and Space Scientists) Compared with Other Scientists, 1970

FIGURE 1   Age Distribution of Professional Meteorologists. Source: Miles F. Harris, *Opportunities in Meteorology* (New York: Vocational Guidance Materials, in Association with the American Meteorological Society, 1972): 155.

My account reinterprets this era in two ways. First, inspired by studies of scientific practice, I explore the daily work practices of the Weather Bureau: collecting observations, making maps, and issuing regularly scheduled forecasts.[6] In light of their successful experience and the daily demands placed upon them, the hesitancy of Bureau forecasters to embrace alternative theoretical models of meteorology seems quite sensible. Second, I frame the adoption of the Bergen School concepts as part of an ideological transformation driven by aviation interests. Carl-Gustaf Rossby successfully persuaded a series of audiences that dynamic meteorology, the branch of meteorology that dealt with the theory and physics of weather, was the best solution for explaining and predicting the weather phenomena crucial to aviation. Sometimes wrestling with the synoptic meteorologists who ran the Weather Bureau, Rossby and Reichelderfer positioned dynamics at the core of meteorology. As Robert Marc Friedman has shown, dynamic meteorology grew out of 19th-century German physics, brilliantly applied to the atmosphere by Vilhelm Bjerknes and the Bergen School. By positioning dynamical explanations at the heart of meteorology, Rossby encouraged meteorologists to aspire to create a universal science of the atmosphere.

Yet even universal sciences grow from particular contexts. In the wake of World War I, as Deborah Coen shows in this volume, the political and geographical situation of a defeated Austria made specific, local studies the most reasonable path to a successful future for the Central Institute for Meteorology in Vienna. The world looked different to the War's victors. The United States emerged from World War I as a recognized world power, undamaged by the fighting, with the strongest economy of any nation. With a unified, continental-scale weather reporting network and control over its vast ocean approaches, global meteorology looked possible from the U.S. Rossby saw a career opening and secured an American-Scandinavian Foundation fellowship to escape the depressed economy of his native Sweden in 1925. As Gregory Cushman points out, Jacob Bjerknes came to the U.S. in 1940 in part because the American context offered him the chance to do large-scale science that could not be carried out in Europe. But it was the American enthusiasm for aviation, another result of World War I, which offered the strongest support for a universal science of the atmosphere. Air power advocates like Gen. Billy Mitchell and Hap Arnold used American geography—especially the wide seas that separated the U.S. from its potential enemies—to argue for the development of aviation generally, and for building airships and long-range, heavy bombers in particular. Lt. Reichelderfer led the Navy Aerology section during the 1920s and 1930s, working closely with the Navy's dirigible program. Reichelderfer and Mitchell realized early in the 1920s that long-range bombers and airships needed meteorological support of a different kind than had evolved in the Weather Bureau over the previous fifty years. High-level winds, the structure of squall lines, and icing conditions in the upper air were irrelevant to a farmer or produce shipper, but of life or death importance to a flyer.

If aviation was the primary source of political and economic support for a universal meteorology, institution building and pedagogy were the mechanisms by which it was created. As Sharon Traweek has pointed out, modern scientific communities are replenished through education.[7] While existing Weather Bureau meteorologists tended to be allergic to the math and physics of Scandinavian dynamic meteorology, a new generation of meteorologists *learned* to aspire to a universal science of the atmosphere from teachers like Rossby and Jacob Bjerknes. These men and their allies then created places where their students could work. The intimate experience of education instilled the universal ideal. David Kaiser's broad interpretation of pedagogy, which expands beyond classroom techniques to encompass the various institutions of training, illuminates how disciplinary cultures are created. Likewise Kenji Ito's work on the movement of the *Kopenhagener Geist* to Japan highlights how educational role-models shape the transmission of values and ideals across national contexts.[8]

This paper begins by describing the culture of the Weather Bureau in the 1920s, exploring how an intensely visual approach to forecasting emerged from daily work practices. The paper next follows the early career of Carl-Gustaf Rossby. Rossby's brief tenure in the Weather Bureau introduced Bergen School concepts into American meteorology, before the ambitious young foreigner issued a forecast for Charles Lindbergh and was fired for transgressing the Bureau's hierarchy. Hired by Harry Guggenheim, the richest aviation booster of the 1920s, Rossby established a model weather service that used Bergen methods to demonstrate that airlines could fly safely and regularly. At the urging of the Navy, Guggenheim then paid for Rossby to become a professor at MIT, where Rossby taught Navy flyers and began to theorize about the general circulation of the atmosphere. The paper returns to the Weather Bureau, focusing on efforts to reform the Bureau after the crash of the Navy dirigible *Akron* in a 1933 squall. After the leadership of the Weather Bureau is firmly in control of the universal meteorologists in 1938, the paper shifts to the training programs established for Army Air Force officers in the lead-up to World War II. The paper explores the education of these "flying fighting weathermen" using the yearbooks they produced to memorialize their experience. Managed by Rossby, the training programs taught a comparatively huge group of meteorologists to see weather forecasting as a problem in applied dynamics. The narrative concludes with a study of Rossby's reforms of the American Meteorological Society in 1944–45, which emphasized the privileged place that theoretical research and academic credentials played in the identity of a new scientific discipline now institutionally and intellectually secure, thanks to the success of the Army Air Force in World War II.

SEEING THE WEATHER: THE US WEATHER BUREAU IN THE 1920S

In the five decades since the establishment of a national weather service in 1871, the US Weather Bureau grew into the dominant American meteorological institution.[9] While a few independent observatories collected weather data, the Weather Bureau played the key role in organizing and administering American meteorology. Nearly all people paid to theorize about or forecast the weather were in some way connected to the Weather Bureau. Even as late as 1940, the Weather Bureau employed over two-thirds of all the trained or experienced meteorologists in America. The Bureau controlled the observational network and the nation's climatology records. University graduate programs in meteorology were basically non-existent, and few other institutions played a role in meteorology.[10]

Within American meteorology, only the Weather Bureau had a significant and reliable source of funds. While congressional appropriations were never adequate to meet the Bureau's perceived needs, and became particularly inadequate as aviators required more extensive upper air observations during the 1930s, the Bureau's budget towered above all the other sources of funding for meteorology in the interwar period. In 1923, the Bureau's appropriation was a bit over $1.9 million, while by 1932, it had swelled to just under $4.5 million. In comparison, the creation of a new meteorology graduate program at MIT in 1928 depended upon a $34,000 grant from the Guggenheim Fund.[11] On the public stage, the Weather Bureau was nearly synonymous with orthodox meteorology. The title of a 1920 children's book suggests how inseparable weather science was from the organization that controlled it: "Gilbert Weather Bureau (Meteorology) For Boys."[12] The Bureau issued the forecasts that appeared in newspapers. It was the first authority journalists consulted when judging the validity of meteorological claims.[13]

The Bureau upheld orthodox meteorology through a culture of conservatism. A domain as important—and public—as the weather attracted many people who claimed revolutionary techniques for improving forecasts or controlling the weather. In terms of social standing, these claimants ranged from fast-talking, itinerant rain-makers like Charles M. Hatfield to Dr. Charles Greely Abbott, a student of astrophysical connections to weather—and the assistant secretary of the Smithsonian Institution. These techniques almost never proved effective or repeatable, but neither did they die. By attracting the attention of a powerful official or two, nearly any idea about the weather might gain celebrity. In 1934, the Secretary of Agriculture appointed a statistician to investigate astrometeorological connections for long-range forecasting, an idea long rejected by the Bureau.[14] The Bureau maintained its intellectual and political authority by generally refusing to support meteorological claims made by people outside of the organization. Bureau officials expected that new techniques would gain only brief, though bright and annoying, prominence, and then fade into the regular background of extravagant stories invented by kooks.

The Bureau's daily forecasting practice depended upon tried and true techniques that had developed alongside the synoptic observing network. According to historian Donald Whitnah, "the general forecasts of 1933 did not vary basically from those of 1871. The movements and relationships among areas of high and low barometric pressure formed the primary source of Weather Bureau prognostications."[15] Synoptic weather maps revealed those pressure area movements. The daily production and interpretation of the synoptic map formed the base of a Bureau man's knowledge of weather. As a torrent of weather observations flooded over the telegraph lines from remote observing stations each morning, meteorologists at the forecasting stations plotted the synoptic weather map. After plotting work by junior figures, the station's "meteorologist-in-charge," usually the most senior forecaster, would analyze the chart and dictate a forecast. Observing stations were widely dispersed, however, and a useful weather map depended upon the continuities of barometric pressure areas, of isothermal lines and bands of precipitation. In creating and interpreting weather maps, forecasters routinely used interpolation and educated intuition. This practical knowledge was born out of experience and the feedback of a daily routine of watching the play of weather observations across a map of America.[16]

The Bureau's leading forecasters attempted to codify their practical knowledge in 1916. Initiated at the request of the new chief Charles F. Marvin, *Weather Forecasting in the United States* attempted to "explain, more or less fully and in detail, the processes by which forecasts can be made," intended for the "guidance and instruction of beginners."[17] But forecasting was an activity that was learned by doing and watching, not reading. "The consensus of opinion seems to be that the only road to successful forecasting lies in the patient and consistent study of daily weather maps."[18] Emphasizing that point, more than one hundred weather maps illustrated 370 pages. The book was a supplement to apprenticeship, and made sense primarily in that context.[19]

The Weather Bureau's recruitment and training practices reinforced the development of forecasting through practice. Lead forecasters typically worked their way up from Junior Observer after entering the Weather Bureau with a high school education.[20] New weather forecasters were promoted from the ranks, "by choosing the winners of contests in making daily practice forecasts."[21] With little formal education or opportunities in meteorology available outside the Weather Bureau, seniority—rather than education or research attainments—largely determined status. Keeping one's intellectual and social distance from unorthodox theories of forecasting (whose advantages were usually illusory) demonstrated the sobriety that distinguished a reliable weatherman. Successful careers were made by observing instruments carefully, plotting maps accurately, and forecasting responsibly. In 1925, an effervescent Swedish visitor stepped into this culture.[22]

BERGEN COMES TO AMERICA

Carl-Gustaf Rossby was 26 when he arrived in America. Rossby was initially appointed by Chief Marvin as a research associate to confer with the Bureau's forecasters and "demonstrate the application of the Bjerknes method of weather forecasting" in the United States.[23] The "Bjerknes method" is better known today as the concepts of the Bergen School. Emerging from the work of the meteorologists assembled by Vilhelm Bjerknes in Bergen, Norway, this approach to meteorology introduced new concepts for analyzing weather maps. The most distinctive of these concepts were the polar front and air mass analysis. Conceived during the waning years of World War I, the polar front represented the boundary between cold air coming down from the north and warm, tropical air moving toward the pole. Where these armies of air collided, storms emerged. Vilhelm Bjerknes's son, Jacob, proposed a novel model of cyclogenesis (storm-creation) in 1919–1920, which emphasized the three-dimensional physical nature of storms. Later in the early

1920s, these concepts were extended through air mass analysis. While just one polar front could exist, air masses could explain the formation of storms at multiple locations. Sometimes thousands of miles across, large pockets of air took on the characteristics of their surroundings if left undisturbed for a few days. Maritime tropical air became warm and moist, while a continental polar air mass was cold and dry. When these air masses bumped against each other, the thinking of the Bergen school went, the physical laws of hydro- and thermodynamics could be used to explain how they would interact. Connecting dynamic and synoptic meteorology, using physical explanations to improve weather forecasting, was a characteristic desire of the Bergen School.[24]

The desire to connect dynamic and synoptic meteorology grew out of Vilhelm Bjerknes's attempt to build a career in classical physics on the periphery of the European scientific community at the end of the 19th century. By connecting storms and weather to theory, Bjerknes aimed to "appropriate the weather" for physics, in Robert Marc Friedman's phrase. The desire to reformulate meteorology as a branch of physics and convert existing meteorologists to this vision marked many Bergenites. Desirable allies were, first and foremost intelligent, rather than kind, experienced or senior. Understanding the physical reasoning behind the techniques mattered most. By the early 1920s, Bergenites were thinking of themselves as "apostles" of the polar front.[25]

Just a year after arriving in Bergen, Carl-Gustaf Rossby became one of these apostles. At Bergen, Rossby had developed twin reputations for charm and brilliance. One of Rossby's co-workers, Tor Bergeron, remembered Rossby's "budding eloquence and power to persuade people to do the things they least of all had intended to do." The young man's "far-reaching ideas and high-flying plans often took our breath away."[26] Rossby returned to his native Sweden to preach to the Swedish meteorological service in 1922.[27] In the following years, Rossby also earned an advanced degree in mathematical physics from the University of Stockholm. By 1925, Rossby set his sights on broader horizons. He won a fellowship from the American-Scandinavian Foundation, promising to study dynamic meteorology problems and the application of the polar front to American weather forecasting.[28] Rossby's experience at the Weather Bureau seems to have begun well. Though institutional culture and Congressional disapproval had made research a low priority at the Bureau, Chief Marvin lauded Rossby's character and intellect in a letter to the American-Scandinavian Foundation in October, 1926.[29]

More importantly, Rossby made an enduring friendship with Lt. Francis Reichelderfer, an officer who headed the U.S. Navy's Aerology section. Young men with a strong scientific education (Reichelderfer had worked as a chemist before World War I), they both worked outside the stable hierarchy of the Weather Bureau. Reichelderfer had become interested in the Bergen methods following a near-disaster in 1921. Flying as an observer during an air power demonstration, Reichelderfer's plane was badly shaken by an unpredicted squall line. This storm front forced the demonstration's leader, General Billy Mitchell, to land his plane on the beach to avoid a crash. Mitchell presented this demonstration, a simulated attack on the captured German battleship *Ostfriesland*, as evidence of the superiority of aircraft to battleships. For Reichelderfer, it was a spur to explore the Bergen methods.[30]

As the head of the weather services for Navy aviation, Reichelderfer had extensive contacts in the aviation community. The most valuable was Harry F. Guggenheim, an heir to the famous mining fortune. Guggenheim had learned to fly in the Navy during World War I. He met Reichelderfer through balloon racing, a sport "as stylish as polo or yacht racing" in the early 1920s; the Navy often fielded entries.[31] Guggenheim believed aviation would be essential to economic and military power in the future, and believed the U.S. was falling behind Europe. As a remedy, he convinced his father to endow The Guggenheim Fund for the Promotion of Aeronautics in 1925.[32] Like most early aviators, Guggenheim appreciated the importance of weather forecasting. Reichelderfer persuaded him to support Rossby's research after the American-Scandinavian fellowship ran out.[33] In August 1927, Guggenheim appointed Rossby chair of the Fund's Committee on Aeronautical Meteorology.[34]

Rossby's connections to the Guggenheim Fund soon brought him into trouble with the Weather Bureau, however. Playing the apostle, he worked to persuade Weather Bureau forecasters to adopt Bergen techniques. "Unfortunately," notes meteorologist-turned-historian Charles Bates, "Rossby was lecturing staid bureaucrats 40 yr his senior."[35] Rossby's strained status with the Bureau leadership took a fatal turn when he made an unauthorized forecast for Charles Lindbergh. Following his famous crossing of the Atlantic, Lindbergh had toured the country sponsored by the Guggenheim Fund. When Lindbergh decided to make a 27-hour winter flight from Washington D.C. to Mexico City, he ignored the Weather Bureau and went straight to Rossby. Public forecasts were strictly the dominion of Weather Bureau regulars. An incensed Chief Marvin allegedly declared Rossby *persona non grata*.[36]

THE GUGGENHEIM MODEL AIRLINE (1927–1928)

Expelled from the Weather Bureau, Rossby began the organizing activities that would eventually make him the leader of the American meteorological community. But at the time, he was a simply an ambitious young foreigner with charm and connections. Harry Guggenheim put Rossby to work developing meteorological services in support of the Western Air Express, a model airline the fund was supporting in California. Rossby's "experimental weather reporting service" would forecast upper air conditions, cloud cover, head- and tail-winds, and landing conditions at airfields between Los Angeles and the San Francisco Bay area.

Rossby spent June and July 1928 establishing observing stations and learning the airways between San Francisco and Los Angeles. His expense reports include bills for 1000 weather maps, $80.00 for a pilot to fly him up and down California for ten days, and $19.25 for a leather flying helmet and goggles.[37] Each observing station was connected by telephone to centralized collecting offices in Oakland or Los Angeles. In addition to air mass analysis and fronts, Rossby's operation drew upon the Bergen school's use of cloud forms to inform prediction.[38] Observers were given a cloud atlas and were expected to include cloud formations in their reports.[39] They were explicitly instructed not to conflate cloud movement with surface winds.[40] To obtain adequate upper air measurements, Rossby contacted the Army Air Corps, hoping they might be willing to launch regular flights to take upper air data, and obtained a number of pilot balloons and related equipment from the Navy.[41]

Rossby also had to recruit and train forecasters. While fishing for weather talent in early 1928, Rossby was introduced to Horace Byers, a junior at Berkeley with an interest in physics and climate. By July, Byers was Rossby's trusted lieutenant, on the Guggenheim payroll at $175 per month. According to Rossby's letters to the Fund's headquarters, Byers was doing "splendid work" teaching the Weather Bureau how to run the reporting service, and beginning empirical studies of California weather.[42]

The Experimental Weather Reporting Service proved successful during 1928 and 1929. Western Air Express suffered no weather-related crashes or mishaps, while the efficiency and reliability of its schedule increased. As the Weather Bureau's San Francisco meteorologist-in-charge noted, the Experimental service's forecasts were also utilized by members of the Automobile Club of Southern California and the California State forester.[43] Despite its successes, the model reporting service nearly collapsed in the spring of 1929 when the Guggenheim fund planned to turn the service over to the Weather Bureau. Weather Bureau headquarters refused to fund the service, and only a last-minute action by departing President Calvin Coolidge (thought to be instigated by incoming President Hoover) explicitly funded it.[44]

TEACHING UNIVERSAL METEOROLOGY, PART I: GRADUATE PROGRAMS

Late in 1927, Edward Warner, formerly Guggenheim professor of aeronautics at the Massachusetts Institute of Technology, now assistant secretary of the Navy for Aeronautics, let Harry Guggenheim know that the Navy needed a good training course for its weather forecasters. Fishing for financial support from "some public-spirited citizen or organization like your own," Warner sought a "fully rounded course to prepare men for meteorological work either in the services or in civil life."[45] A year layer, Guggenheim's public spirit brought Rossby to MIT as assistant professor of meteorology.

Rossby began training a group of colleagues in addition to teaching his military students.[46] Horace Byers became the department's first graduate student. Jerome Namias, Harry Wexler, and Athelstan Spilhaus studied at MIT during the 1930s, and all went on to distinguished careers in geophysical research. In the spring of 1929, Hurd C. Willet, a former Weather Bureau observer and one of the handful of Americans with a Ph.D. in meteorology, joined the department as assistant professor.[47] Rossby's charm and generosity marked his teaching. Horace Byers recalled that Rossby's "informal discussions over luncheon or a cup of coffee . . . were nothing less than an inspiration."[48]

Rossby's most important research occurred during the ten years he spent at MIT. This work focused on exploring the influence of upper air conditions upon the movements of air masses. Working from the increasing number of upper air observations being taken around the northern hemisphere (and some of which he arranged in the Boston area), Rossby developed techniques for identifying and tracking air masses. By charting constant potential temperature (isentropy), the boundaries of air masses could be identified and followed over time. To understand what caused the movement of air masses, Rossby returned to the rotation of the earth, and its effect on the general circulation of the atmosphere. From upper air observations, he teased out a pattern of enormous tongues of low pressure, reaching downward from the pole. These large-scale atmospheric disturbances, he realized, were waves that had periods of a few days, the same order as many substantial changes in weather. Following a line of mathematical reasoning that stretched back to Laplace's theory of ocean tides and through a theorem of Hermann von Helmholtz, Rossby developed an equation that could be used to calculate the movement of these waves. This equation, published in 1939, accounted for shifts in the upper-level westerlies, the steering winds which guided air masses. The result made Rossby world-famous amongst physicists studying the atmosphere.[49]

While Rossby was making an international reputation, aviation continued to grow, despite the global economic collapse. Between 1928 and 1940, five American universities established graduate programs in meteorology. Following MIT in 1928, the California Institute of Technology established a program in 1933,[50] and Rossby's student Athelstan Spilhaus led a department at New York University from 1937. These programs were each connected to Guggenheim-funded schools of aeronautics. In 1940, the University of Chicago and the University of California, Los Angeles introduced professional meteorological instruction. All five programs focused upon preparing students for careers in aviation, and began to send graduates into the Weather Bureau following reforms in the 1930s.

REFORMING THE WEATHER BUREAU

While Rossby worked at MIT, flying interests called for improvements in the Weather Bureau's aviation forecasting services. The resulting reforms embraced the values of the Bergen School. By the time Rossby rejoined the Weather Bureau in 1938 as assistant director for research, directives from outside the Bureau had set it on a course towards full adoption of Norwegian methods. University education of young men was the primary mechanism for spreading the Bergen School.

The Air Commerce Act of 1926 directed the Weather Bureau to provide forecasts and warnings useful for aviation. Flying grew far faster than appropriations, however, and the Bureau struggled to provide adequate observations and forecasts for the nation's airways. Nor did the economic collapse of 1929 help appropriations, though it was not until the first Roosevelt budget that government spending was drastically curtailed. The Weather Bureau lost nearly $2 million, about 45% of its budget, between the 1932 and 1933 appropriations. Chief Marvin dismissed nearly 500 employees, about 20% of the Bureau's workforce.[51]

On April 4th, 1933, the Navy dirigible USS *Akron* crashed in an unpredicted squall, killing 73, including the head of the Navy's Bureau of Aeronautics, Adm. William Moffett. Resulting Congressional hearings explored the inadequacies of the Weather Bureau, while highlighting the credibility problems faced by such a public science. In one hearing, Charles Mitchell, the Bureau's most respected forecaster, battled a Senator who claimed expertise based upon many years of studying Weather Bureau reports, while the committee chairman spoke of meteorology as "this so-called 'science.' " Later in the hearings, air power advocate Billy Mitchell argued that the weather service should be rearranged, removed from the Department of Agriculture (which was responsible for "raising onions, potatoes and such things,") and run by the military.[52]

President Roosevelt took a moderate response, calling upon a new institution: the Science Advisory Board.[53] Chaired by MIT president Karl Compton, the Board appointed a subcommittee to recommend improvements to the Weather Bureau: one Weather Bureau meteorologist, Charles D. Reed, and three university presidents: Compton, Johns Hopkins's Isaiah Bowman, and Cal Tech's Robert Millikan. In addition to being America's most distinguished physicist, Millikan had headed the Army's meteorology program during the Great War.

The subcommittee's report called for the immediate adoption of "air-mass analysis methods." With air-mass analysis, the committee wrote, "there is the practical certainty that our whole forecasting service can be improved both as to accuracy and reliability."[54] Since air mass analysis required upper air data, the report suggested that the Army and Navy provide daily aerological measurements during their regular training flights. The report called for the Bureau to keep up with new developments in air-mass analysis. In addition, the Committee recommended that a system of postgraduate training for Weather Bureau meteorologists be initiated. All forecasters should receive "thorough instruction in the more modern methods," while those "who already have a good basic training in meteorology, physics and mathematics and have shown some proficiency in the actual art of forecasting" should be detailed "to an institution of recognized leadership in this field" for six months or a year of advanced instruction.[55] Superior meteorological knowledge now came from universities, this group of college presidents implied.

The Weather Bureau's adaptation to air-mass analysis, however, "proved to be a painfully slow process."[56] The *Akron* crash ended Chief Marvin's tenure in January 1934; he officially retired eight months later after fifty

years with the weather service. The new chief was Willis Gregg, head of the
Bureau's aerology division. Gregg's promotion made clear the centrality of
aviation interests. But while Gregg had been a member of the Guggenheim
committee on meteorology alongside Rossby, he remained an insider who
had spent his career with the Bureau. Between 1935 and 1938, the Bureau
appointed roughly "a dozen young university men with air mass training."[57]
Among these men were Rossby's students Horace Byers and Harry Wexler.
Gregg's choices about how to pay for the new technique may have caused
some of the conversion pains. Money intended for salary increases in 1935
was diverted to the new air mass section instead.

George R. Stewart's best-selling 1941 novel, *Storm*, suggests the tension
within the Bureau between young university men and senior, experienced
forecasters:

> [The Junior Meteorologist] admitted that he had been unhappy in
> the Weather Bureau; his mathematical training did not seem to help
> him, and sometimes he thought that it even was a handicap. Some-
> times it seemed as if the Chief were only using the same methods
> that any shepherd might have used back in the time of the Patri-
> archs; he just looked at the sky, and decided from the appearance of
> things what weather would come along after a while. The Shepherd,
> of course, never saw farther than the actual horizon. By the weather
> map the Chief extended his view for several thousands of miles.
> There was a tremendous pyramiding of information, but not much
> change in method.[58]

To the young men trained in the methods of the Bergen School, the empirical methods of senior forecasters simply looked obsolete. These men seemed
to be stubborn old-timers resisting the progress of modern meteorology.
Reminiscing in 1981, Jerome Namias took the resistance to Bergen methods
as evidence of the Bureau's "provincial and narrow view."[59]

To the experienced forecasters responsible for issuing reliable forecasts
on tight schedules, however, the intuitive processes refined through years of
practice represented a skilled judgment that couldn't be learned in a classroom. Running a weather service was different from talking about one.

The Bergen School learned this difference when Chief Gregg died from
an unexpected heart attack in 1938. At the suggestion of the Science Advisory Board, President Roosevelt appointed Francis Reichelderfer. Commander Reichelderfer learned of the offer as he entered port aboard the *USS*

Utah.[60] Not only was the 43-year-old Navy Officer the first outsider to head the Bureau, he hired Rossby as Assistant Director for research and education.

Reichelderfer proved a conscientious administrator, and he avoided making wholesale changes to his new organization. The Civil Aeronautics Act of 1938 gave him a useful tool for gradual change: education. The act directed the Weather Bureau to send men each year to universities for "training at Government expense . . . in advanced methods of meteorological science." Such men also retained their seniority as they learned.[61] Reichelderfer further reformed the Bureau by addition. Senior forecasters could continue to forecast as they had, while new methods and new men were added to the process.

The Department of Agriculture's *1941 Yearbook of Agriculture: Climate and Man* illustrates these processes of reform. In an early chapter, Reichelderfer laid out the new goals that marked modern meteorology. In its current stage, expert forecasting remained "a combination of training, experience and native ability. However, as progress is made in three-dimensional analysis of the weather and . . . knowledge of its physical processes . . . the science will become more systematic and exact." Progress would come through research. Meteorology's goal was not only better forecasting, but also to diminish the importance of "personal factors" in the process.[62]

Rossby's contribution reiterated the commitment to research as the engine of progress. In "The Scientific Basis of Modern Meteorology," Rossby laid out a "semi-technical" presentation of the physics of Northern hemisphere atmospheric circulation. "Genetics, soil science and nutrition have all made great strides based upon important fundamental discoveries," he argued. "Latest to join this group is meteorology."[63] Arguing for the primacy of theory, he wrote, "it is safe to say that until the proper theoretical tools are available, no adequate progress will be made either with the problem of long-range forecasting or with the interpretation of past climatic fluctuations."[64] While much of his article seems far too "semi-technical" for the average reader, he ended his piece with a short demonstration of "amateur forecasting from cloud formations."

Rossby's heart didn't seem to be in popularization or administration, however. In 1941, he followed his protégé Horace Byers to the University of Chicago, where Rossby took over a new Institute for Meteorology. From the Institute, Rossby organized academic meteorology through the University Meteorological Committee, which eventually supplied thousands of weather forecasters to support an air force capable of winning a global war.

TEACHING UNIVERSAL METEOROLOGY, PART II:
FLYING FIGHTING WEATHERMEN

Franklin Roosevelt's 1940 call for a 50,000-plane air force set Army and Navy officers to thinking about training. New squadrons would need pilots and bombardiers—and weather officers. War would also require establishing weather-observing stations around the globe.[55]

In 1940, the few dozen existing military weather officers had largely learned their meteorology from MIT or Cal Tech during the 1930s. They looked to the research universities for expanded training programs. As of October 1940, there were 150 new cadets studying meteorology at MIT, Cal Tech, NYU, UCLA, and the University of Chicago.[66] Following the attack on Pearl Harbor, the immediate need for new meteorologists loomed large. As the five schools went to year-round programs and began taking in two classes of meteorological cadets per year in 1942, Col. Donald Zimmerman, the head of AAF Weather Service, estimated that the AAF would need 10,000 weather officers by the start of 1945.[67]

The University Meteorological Committee (UMC) saw itself as "a clearing house for the exchange of ideas between the [AAF] Directorate of Weather and the individual universities," while the committee was "at the disposal of the Army Air Forces whenever technical problems arise."[68] The UMC coordinated the training program, exchanging research, curricula and even instructors.

The UMC had to secure an adequate number of suitable young men to train.[69] Advertising and recruiting efforts were extensive, even including a half-hour radio program titled "The Invisible Allies! A Thrilling Chapter from the Notebooks of Science—and War!" that featured an appeal from Nobel laureate physicist Arthur Compton.[70] The graduate program's stiff educational prerequisites (one year of college physics, differential and integral calculus, and the successful completion of two full years of college) made necessary a second and third series of preparation programs.[71]

The curriculum mixed a heavy dose of physical theory with hundreds of hours of synoptic map analysis:

"The program offered by the Institute for the training of cadets as meteorologists for the Army Air Corps, and of Navy officers as aerologists, has been planned to provide each student with the utmost practical experience in the analysis of weather charts and forecasting,

157

and at the same time to endow him with a broad theoretical background in modern meteorological physics. To accomplish this, the academic week at the Institute comprises approximately 20 to 24 hours of practical weather analysis and forecasting in the synoptic laboratories and 12 hours of formal lectures. From two to four hours of the weekly laboratory time are devoted to discussions, . . . A two-hour examination covering all subjects is held weekly, and final examinations are given in each course."[72]

While this pattern adhered to the academic mold, cadets also learned to march, shoot, salute, and defend themselves against poison gas attack when they were not calculating radiative cooling.[73] The students called it a "GI life of calculus, physics, and meteorology."[74]

The Bergen School's abstract way of knowing weather formed the core of the curriculum. The textbooks taught that the basic goal of meteorology was to understand weather in physical and mathematical terms, so it could

Curriculum for Wartime Training Classes
Institute of Meteorology, University of Chicago

Subject	Total # of hours
Synoptic laboratory	672
Dynamic meteorology	116
Synoptic and aeronautical meteorology	72
Introductory meteorology	43
Hydrodynamics	42
Field course	24
Radiosonde	53
Climatology	32
Geography	22
Physics of the High Atmosphere	32
Oceanography	20
Fieldwork with mobile weather unit	18
Examinations	95

Books Issued to Meteorology Cadets, UCLA A-level Class, March 1943[75]

Horace R. Byers	*Synoptic and Aeronautical Meteorology*	McGraw-Hill, 1937
Bernhard Haurwitz	*Dynamic Meteorology*	McGraw-Hill, 1941
Bernhard Haurwitz	*The Physical State of the Upper Atmosphere*[76]	Royal Astronomical Society of Canada, 1941
Wilfred Kendrew	*The Climates of the Continents,* 3rd edition	Oxford University Press, 1937
W. E. K. Middleton	*Meteorological Instruments*	University of Toronto Press, 1942
Jerome Namias	*Air Mass and Isentropic Analysis*	American Meteorological Society
Sverre Petterssen	*Weather Forecasting and Analysis*	McGraw-Hill, 1940
Athelstan Spilhaus and James Miller	*Workbook in Meteorology*	McGraw-Hill, 1942
Victor Starr	*Basic Principles of Weather Forecasting*	Harper, 1942
Harald Sverdrup	*Oceanography for Meteorologists*	Prentice-Hall, 1942
G. E. F. Sherwood and Angus Taylor	*Calculus*	Prentice-Hall, 1942
Weather Bureau	Circular N	

be quantified and calculated.[77] Victor Starr's *Basic Principles of Weather Forecasting* described numerical calculation as the ultimate goal, while admitting that it remained distant. "Since we know the fundamental laws" of fluid mechanics and thermodynamics, "it might appear that a forecast of future motions could be made completely by analytical means. Unfortunately, although we do know the elementary principles of atmospheric motions, the problem of integrating them and obtaining a forecast by purely analytical procedures is too complex to be treated by a direct frontal attack." Starr then footnoted Lewis Fry Richardson's 1922 "effort in this direction."[78]

Weather was generally presented as a secondary phenomenon, the local consequences of the general circulation of the atmosphere. "The general problem of forecasting weather conditions may be subdivided conveniently into two parts. In the first place, it is necessary to predict the state of motion of the atmosphere in the future," wrote Starr. As a second step, only after predicting the atmosphere's motion, "it is necessary to interpret this expected state of motion in terms of the actual weather which it will produce at various localities. The first of these problems is essentially of a dynamic nature, inasmuch as it concerns itself with the mechanics of the motion of a fluid." Successful forecasting resulted from first understanding the physical principles that governed the atmosphere.[79]

While reforming meteorology around a quantitative and physical understanding of the atmosphere was central to Rossby and the Bergen School's long-term agenda, the military needed to be convinced why it should pay for (and wait for) its cadets to learn so much theory. Rossby drew upon the claims of physics to universality:

> Earlier methods of training meteorologists, particularly in the United States Weather Bureau, were based entirely on the accumulation of experience. A man trained over a number of years in, say, San Francisco, would in that fashion become a good forecaster for our West Coast but would have to start all over again if he were transferred to another part of the country.
>
> We do not have the time to give our students adequate basic training and also a large amount of experience within the short period of time at our disposal. Hence, we *must* concentrate on the application of fundamental principles of analysis and forecasting which can be used in any part of the world.[80]

Rossby argued that meteorologists grounded in such principles were not only superior, but also the only effective kind that could be produced quickly. Dynamic meteorology promised portable, placeless knowledge. Knowing the weather through physics, Rossby argued, meant forecasters could work effectively anywhere the global war might require them.

Wartime experience in the tropics suggested otherwise, however. New forecasters found that techniques developed in Norway did not always describe tropical weather patterns. In the South Pacific and Latin America, experienced dispatchers and forecasters from Pan American Airlines provided the local knowledge necessary to safely route planes.[81] The UMC established an Institute of Tropical Meteorology at the University of Puerto Rico to train advanced students—and conduct research to expand dynamic meteorology into tropical skies.

While the UMC curriculum taught students to understand the weather in terms of physics, other aspects of the wartime experience shaped the emerging culture of modern meteorology. First and foremost, the weather cadets were military men, under the orders and authority of the Army Air Force.[82] When the cadets graduated, they earned officer's bars, and they were expected to be leaders and soldiers. The UMC classes were powerful selectors on the basis of race, sex and class. Very few black men became weather officers, and those who did were segregated to support all-black fighter units.[83] About two hundred women studied meteorology in the universities during the war.[84] The high educational prerequisites almost certainly excluded people from poorer and immigrant backgrounds. These excluded groups missed the emerging social and educational networks so central to postwar professional meteorology.

THE CADET EXPERIENCE

Although the students were military officers in training, the meteorology training program retained a collegiate atmosphere. College professors led lectures and labs, while students took exams and lived together in dormitories or apartments. Like many college classes, the weather cadets often produced commemorative yearbooks to construct and solidify the meanings of their experiences. These books followed the genre conventions of other memorial yearbooks.[85] Full of inside jokes, the student authors drew upon military and weather metaphors to recreate shared experiences. For example, "the big guns—Bjerknes and Kaplan and Holmboe—bombarded us without quarter," and "the worst maps turned up as test maps, which is what is meant by periodicity in weather."[86] These books offer a way inside the classroom to see how meteorologists-in-training understood their education.

Reflective writers in the meteorology classes felt their training instilled a new way to know and understand weather. Entering the military, the cadets understood weather as something experienced bodily and discussed in commonplace terms. Learning to see their surrounding environment in more abstract ways de-centered their individual experiences. Personal feelings, bodies, and the particulars of place became increasingly extraneous to the description of events:

> The world around us changed quickly; a cloudy sky became a nimbostratus overcast, the wind that tugged at our overcoats became a

Beaufort 6 and we all looked for 00 Wx [clear weather] on our weekends. The days of our years took on a more universal aspect and almost unconsciously we came to measure time by new and different standards. . . . The doldrums moved South despite the fact that 150 neophytes had been strapped to the NYU assembly belt and were being processed into weathermen.[87]

The students felt the tension between the vernacular weather culture they entered with and the scientific way of seeing they were being taught. While they felt a chilly wind tugging at an overcoat, they learned to abstract that gust into a number on the Beaufort scale that could be compared to other wind measurements anywhere in the world.

Yet for the bomber crews the weather cadets were training to serve, weather remained a phenomena experienced in the gut and the fingers. Some weather cadets flew training missions to learn how to forecast and surveil weather from the air.

In reward for beavering well done, we were given an opportunity during that last quarter to fulfill our manifest destiny—that of becoming "flying fighting meteorologists." A convalescent B-18 was assigned to our detachment for weather reconnaissance, and group by group we groundlings took to the air verifying our own forecasts. Frozen limbs and upset stomachs made lapse rates and turbulence less academic, and we returned sadder but wiser weathermen.[88]

Despite the urgings of their academic teachers towards theory, abstraction, and the rejection of subjective experience, the irreducible world of embodied experience remained a fundamental category for fully understanding the meanings of their new role. To be a true fighting meteorologist, a man had to both understand the physics of weather, as well as the physical experience of what those map symbols meant to the men flying bombers through limb-freezing, turbulent air against murderous enemies.

PREPARING A POST-WAR DISCIPLINE

By 1944, it was clear that the UMC had produced more than enough new weather forecasters for the AAF. As the training programs wound down, Rossby turned his attention towards shaping the structure of post-war meteorology. As president of the American Meteorological Society, Rossby spent 1944–1945 overseeing a major transformation of the AMS (while also trotting the globe as an Air Force consultant). He reformed the constitution, creating a two-tiered membership structure, and introduced a new technical journal, the *Journal of Meteorology*. He worked to create a placement service for meteorologists, promoting the development of "industrial meteorology." He also encouraged the development of meteorology classes as part of the liberal arts curriculum, while working to integrate meteorology into the work of civil engineers and the training of research geologists, oceanographers, and hydrologists.[89]

Rossby's constitutional reforms at the AMS institutionalized the authority and power of university-trained meteorologists. A June 1944 letter "to the Members of the American Meteorological Society from the Council" reveals this institutionalization. The letter accompanied a ballot for members to vote on significant changes in the bylaws and constitution. Splitting the membership into two castes most clearly stated the Society's new priorities. "It is recognized that the Society contains two broad groups of members," the Council wrote performatively, "one consisting of those employed as professional meteorologists, and the other consisting of sub-professional and interested amateurs." Though the two groups have different needs, "it is clear that the future strength of American meteorology lies in unity rather than in independent action by separate groups."[90] Instead of schism, putting the professionals in charge would strengthen American meteorology. "Since it was felt that the administration of the Society should be in the hands of those most vitally interested in the science, it is proposed that the President, Vice President, Secretary, and three of the five Councilors elected each year be Professional Members." Professional members would also be charged $10.00 per year, a hefty increase over the rate of regular membership, $3.50 in 1943.

By giving control over the Society to those experts with either education or employment in meteorology, the constitutional change created an organization dedicated primarily to the interests and concerns of its professional members, while retaining the potential political influence that comes with a large membership. It also helped to move meteorology further away from the public, a useful move for a science long plagued by credibility problems.[91] Twenty-five years later, Horace Byers applauded the stratification of the AMS: "Professional membership now distinguished the trained and experienced meteorologists from the hacks and the dilettantes, and the Society was taking its place among the distinguished learned societies."[92]

In addition to changing the membership structure, Rossby oversaw the introduction of the *Journal of Meteorology*, a new journal dedicated to publishing high-quality theoretical research, including research that had been conducted during the war as it was released from security restrictions. The journal became an important organ for research into the atmosphere's general circulation, and featured Seymour Hess's study of Mars in 1950. Finally, Rossby initiated a system of personnel cards, intended for use as a placement service in peacetime, and for quickly locating properly trained meteorologists during future wars. Extensive records in the UMC files show Rossby's attempts to encourage various sectors of private industry to explore how meteorologists might be of value. From the pulpit of the AMS presidency, he also encouraged the integration of meteorology into the general liberal arts curriculum, thereby hoping to create a need for meteorology teachers at the high school and college level.

CONCLUSION: A THRIVING COLD WAR DISCIPLINE

American meteorology emerged from World War II as a thriving scientific discipline. The aspiration towards a universal science of the atmosphere generated journals, departments, graduate students, national research centers, military support, and research dollars.[93] Global meteorology was also becoming a key element in U.S. foreign policy. In the next essay, Gregory Cushman shows how Rossby and Jacob Bjerknes worked to spread their meteorological approach into Canada and Latin America during the war, and how the U.S. government used the weather cadet training program as a way to orient Latin American meteorology towards the United States. After the war, the State Department promoted "scientific internationalism," to challenge the international appeal of communism. The World Meteorological Organization, created in 1951 with Francis Reichelderfer as its first president, worked to integrate the weather observing systems of member nations into a global network for freely exchanging data.[94]

These efforts coincided with the aspirations to create a universal science of the atmosphere that the weather cadets had learned during the war. The mathematics and physics training at the heart of their education helped them to integrate new technologies for remote sensing and computation into meteorological practice. Radar networks and weather satellites enabled meteorologists to construct synthetic views of the globe, while atmospheric modelers constructed virtual global atmospheres from equations and electrons. Large-scale atmospheric models became the central basis for weather forecasting practice in the later third of the 20th century. Built into these models was the new social order of meteorology: dynamicists created, synopticians applied. Horace Byers pointed this out to the National Academy of Sciences in 1955 when he celebrated Jule Charney's advances in the numerical simulation of atmospheric flows while noting that "a second step, such as pinpointing cloud, rain, and temperature areas, is left for the harassed local weatherman."[95]

While increasingly detailed weather simulations elevated the dynamic meteorologist, they also revealed the limits of the universal ideal for meteorology. By the 1970s, computer simulations had led to the discovery of chaos, one of the ways in which mathematical knowledge of complex systems like weather is fundamentally limited. Ironically, a pioneer in this exploration was Edward Lorenz, who first learned meteorology as a weather cadet in the academic department Carl-Gustaf Rossby had founded.[96]

NOTES

*A National Science Foundation Graduate Research Fellowship supported the research and writing of this essay. The opinions expressed here are mine alone, and do not reflect the official position of the NSF.

1. Seymour L. Hess, "Some Aspects of the Meteorology of Mars," *Journal of Meteorology* 7, 1 (February 1950): 1–13.
2. Werner A. Baum, "Seymour L. Hess, 1920–1982," *Bulletin of the American Meteorological Society* 63 (1982): 215.
3. The American Meteorological Society's highest medal is named for Rossby, and the Society bestows an annual award in Reichelderfer's name to honor a provider of meteorological services to the public.
4. Kristine C. Harper's thoroughly-researched, engaging dissertation is the most recent and comprehensive study: "Boundaries of Research: Civilian Leadership, Military Funding, and the International Network Surrounding the Development of Numerical Weather Prediction in the United States" (Ph.D. dissertation, Oregon State University, 2003). John Lewis's fascinating topical studies draw upon a range of little-seen archival sources: "Carl-Gustaf Rossby: A Study in Mentorship," *Bulletin of the American Meteorological Society* 73 (1992): 1425–1438; "Cal Tech's Program in Meteorology: 1933–1948," *Bulletin of the American Meteorological Society* 75 (1994): 69–81; "C.-G. Rossby: Geostrophic Adjustment as an Outgrowth of Modeling the Gulf Stream," *Bulletin of the American Meteorological Society* 77 (1996): 2711–2728; "LeRoy Meisinger,

Part I: Biographical Tribute with an Assessment of His Contributions to Meteorology," *Bulletin of the American Meteorological Society* 76 (1995): 33–45. Frederik Nebeker *Calculating the Weather: Meteorology in the 20ᵗʰ Century* (San Diego: Academic Press, 1995).

5. Charles C. Bates, "The Formative Rossby-Reichelderfer Years in American Meteorology: 1926–1940," *Weather and Forecasting* 4 (1989): 593–603; Horace R. Byers, "Carl-Gustaf Arvid Rossby," *Biographical Memoirs of the National Academy of Sciences* 34 (1960): 248–270; Horace R. Byers, "Carl-Gustaf Rossby, the Organizer," in *The Atmosphere and the Sea in Motion: Scientific Contributions to the Rossby Memorial Volume*, ed. Bert Bolin (New York: The Rockefeller Institute Press, 1959), 56–64; Jerome Namias, "The Early Influence of the Bergen School on Synoptic Meteorology in the United States," *PA-GEOPH* 119 (1981): 491–500; Jerome Namias, "Francis W. Reichelderfer, 1895–1983," *Biographical Memoirs of the National Academy of Sciences* 60 (1991): 272–291.

6. For example, Robert E. Kohler, *Lords of the Fly* (Chicago: University of Chicago Press, 1992); Pierre Bourdieu, *Outline of a Theory of Practice* (Cambridge: Cambridge University Press, 1977); Andrew Pickering, *The Mangle of Practice: Time, Agency and Science* (Chicago: University of Chicago Press, 1995).

7. Sharon Traweek, *Beamtimes and Lifetimes: The World of High Energy Physicists* (Cambridge, MA: Harvard University Press, 1988), Ch. 3.

8. David Kaiser, *Drawing Theories Apart: The Dispersion of Feynman Diagrams in Postwar Physics* (Chicago: University of Chicago Press, 2005). Kenji Ito, "The *Geist* in the Institute: The Production of Quantum Physicists in 1930s Japan," in *Pedagogy and the Practice of Science: Historical and Contemporary Perspectives*, ed. David Kaiser (Cambridge: MIT Press, 2005): 151–183.

9. Harper, "Boundaries of Research," 7

10. Ibid., Ch. 3. Harper describes the limited educational opportunities of the interwar years. See also William A. Koelsch, "From Geo- to Physical Science: Meteorology and the American University, 1919–1945," in *Historical Essays on Meteorology, 1919–1995*, ed. James R. Fleming (Boston: American Meteorological Society, 1996): 511–540.

11. Weather Bureau budget information: Donald Whitnah, *A History of the United States Weather Bureau* (Urbana: University of Illinois Press, 1961): 181. For more on the Guggenheim Fund, see Milton Lomask, *Seed Money: The Guggenheim Story* (New York: Farrar, Straus and Company, 1964): 125.

12. This book was intended to drive sales of the A.C. Gilbert Company's Weather Bureau set, a collection of meteorological instruments. The advertisement at the back of the book encourages boys to "Learn to use the Gilbert Weather Bureau to read weather indications from instruments set up by yourself . . . Your boy friends will listen to you with interest when you explain to them the cause of

storms and how important it is to have a knowledge of climatic disturbances." Alfred C. Gilbert, *Gilbert Weather Bureau (Meteorology) for Boys* (New Haven: The A. C. Gilbert Company, 1920), 85.

13. Others included the Blue Hill observatory. Some teaching and research into weather and climate occurred in geography departments, as at Clark University and by Charles Brooks at Harvard (Koelsch, "From Geo- to Physical Science"). On journalists and the validity of meteorological claims, see Clark C. Spence, *The Rainmakers: American "Pluviculture" to World War II* (Lincoln: University of Nebraska Press, 1980). For more about the Weather Bureau before the transfer to civilian control, see James Rodger Fleming, "Storms, Strikes and Surveillance: The US Army Signal Office, 1861–1891," *Historical Studies in the Physical and Biological Sciences* 30, no. 2 (2000): 315–32.

14. For more on Hatfield and commercial "pluviculture," see: Spence, *The Rainmakers*. Harper describes the 1920s battles over insolation and a 1934 astrometeorology affair: Harper, "Boundaries of Research," 47–50.

15. Whitnah, *History of the Weather Bureau*, 159.

16. Alfred J. Henry et al., *Weather Forecasting in the United States* (Washington: U.S. Department of Agriculture, Weather Bureau, Government Printing Office, 1916).

17. Ibid., 3.

18. Cited in Bates, "Formative Rossby-Reichelderfer Years," 594.

19. Writing in the early 1990s, university-trained, theoretically attuned Jerome Namias felt differently: "*Weather Forecasting in the United States*, written by top forecasters of the U.S. Weather Bureau, contained hundreds of charts and rules for forecasting, empirically derived and completely lacking in interpretation. Many of the rules seemed contradictory. The book was frustrating to read and, though published in 1916, was studied by few. After just a few years, the book was already a relic, only of historical interest." (Namias, "Francis W. Reichelderfer," 275).

20. Bates, "Formative Rossby-Reichelderfer Years." Charles L. Mitchell, one of the Bureau's leading forecasters in the 1930s and 1940s, described his career track when testifying before Congress at the *Akron* crash hearings (Joint Committee to Investigate Dirigible Disasters, *Investigations of Dirigible Disasters: Hearings before a Joint Committee to Investigate Dirigible Disasters*, 1st Session, 1933: 197.) See also the reminiscences of Jack Thompson, an Observer in the Weather Bureau during the late 1920s and 1930s (Jack C. Thompson, "Weather Prediction at the Local Weather Bureau Office as Concepts from the Bergen School Came to the U.S.," *Bulletin of the American Meteorological Society* 66, (1985): 1250–1254.)

21. Whitnah, *History of the Weather Bureau*, 159

22. "Effervescent" comes from Bates, "Formative Rossby-Reichelderfer Years."

23. Charles F. Marvin to James Creese, October 12, 1926. Published in Norman Phillips and Anders Persson, "C.-G. Rossby's Experience and Interest in Weather Forecasting," *Bulletin of the American Meteorological Society* 82, (2001): 2025.

24. Robert Marc Friedman, *Appropriating the Weather: Vilhelm Bjerknes and the Construction of a Modern Meteorology* (Ithaca: Cornell University Press, 1989).

25. Ibid., 195–199.

26. Tor Bergeron, "The Young Carl-Gustaf Rossby," in *The Atmosphere and the Sea in Motion: Scientific Contributions to the Rossby Memorial Volume*, ed. Bert Bolin (New York: The Rockefeller Institute Press, 1959), 52.

27. Friedman, *Appropriating the Weather*, 195. More on Rossby's work with the Swedish meteorological service can be found in Phillips and Persson, "C.-G. Rossby's Experience and Interest in Weather Forecasting."

28. Several useful articles explore Rossby's life, though he has not received the full-length biography his ebullient personality and accomplishments deserve. For biographical information, see Bergeron, "The Young Carl-Gustaf Rossby"; Byers, "Carl-Gustaf Arvid Rossby"; Norman Phillips, "Carl-Gustaf Rossby: His Times, Personality, and Actions," *Bulletin of the American Meteorological Society* 79, (1998): 1097–1112.

29. Charles F. Marvin to James Creese, October 12, 1926, published in Phillips and Persson, "C.-G. Rossby's Experience and Interest in Weather Forecasting."

30. Bates, "Formative Rossby-Reichelderfer Years," 595.

31. Ibid.

32. The Guggenheim Fund aimed to kick-start American aviation growth. It disbursed its capital to various pioneering projects, catalyzing the development of aviation at a crucial period. The fund ceased operations in February 1930, after expending $2.6 million. The best-known projects supported by the fund included the development of instrument flying techniques, enabling pilots to fly safely through fog and cloud; Charles Lindbergh's triumphant national tour; and a "safe airplane" competition, which aimed to create an airplane as safe and easy to use as an automobile. The fund also endowed schools of aeronautics at Cal Tech, MIT, Stanford, Michigan, Georgia Tech, and the University of Washington (Richard P. Hallion, *Legacy of Flight: The Guggenheim Contribution to Aviation* (Seattle: University of Washington Press, 1977), pp 92, 169–172.)

33. Bates, "Formative Rossby-Reichelderfer Years," 596.

34. Hallion, *Legacy of Flight*, 92.

35. Bates, "Formative Rossby-Reichelderfer Years," 594–5.

36. Ibid., 594–5; Byers, "Carl-Gustaf Rossby, the Organizer," 56. While Byers was eager to dramatize Rossby's banishment to other meteorologists in a 1959 memorial, he chose not to mention it in Rossby's biographical memoir for the National Academy of Sciences (Byers, "Carl-Gustaf Arvid Rossby."). I have not seen contemporary documentary evidence regarding Rossby's exile.

37. Carl-Gustaf Rossby, Expense report, July 1928. Collected Papers of the Daniel Guggenheim Fund for the Promotion of Aeronautics, Library of Congress. Box 8, Files 1 and 2 (Henceforth, DGFPA).

38. For more on the Bergen School, see Friedman, *Appropriating the Weather*.

39. "Cloud Forms, According to the International System of Classification," Published by the United States Department of Agriculture and the US Weather Bureau; Prepared by the Weather Bureau's Cloud Committee. Undated.

40. "Instructions for Observers," DGFPA.

41. Rossby to Maj. G.C. Brant, 18 Aug. 1928. DGFPA.

42. CGR to HFG, 7/23/28. DGFPA.

43. Edward H. Bowie, *Weather and the Airplane: A Study of the Model Weather Reporting Service Over the California Airway* (New York: The Daniel Guggenheim Fund for the Promotion of Aeronautics, 1929), 25–27.

44. Bowie, *Weather and the Airplane*, 24.

45. Edward P. Warner to Harry F. Guggenheim, 3 Feb. 1928. Also see Edward P. Warner to Harry F. Guggenheim, 7 Dec. 1927. DGFPA.

46. John M. Lewis, "Carl-Gustaf Rossby: A Study in Mentorship."

47. Harper, "Boundaries of Research," 112.

48. Byers, "Carl-Gustaf Arvid Rossby," 255.

49. Lionel Pandolfo, "Rossby Waves," in *Encyclopedia of Climate and Weather*, ed. Stephen H. Schneider (New York and Oxford: Oxford University Press, 1996); Carl-Gustaf Rossby et al., "Relation between Variations in the Intensity of the Zonal Circulation of the Atmosphere and the Displacements of the Semi-Permanent Pressure Systems," *Journal of Marine Research* 2 (1939): 38–55.

50. Led by Irving Krick, Cal Tech's program was quite different from the Bergen programs at the other four universities. Despite being Horace Byers's brother-in-law, Krick was hated by many meteorologists. By the 1950s, his support for unorthodox methods and bold forecasting claims made him a boundary case for acceptable professional behavior. For more on Cal Tech's program, see J. M. Lewis, "Cal Tech's Program in Meteorology: 1933–1948," *Bulletin of the American Meteorological Society* 75 (1994): 69–81.

51. Bates, "Formative Rossby-Reichelderfer Years," 597; Whitnah, *History of the Weather Bureau*, 21, 183.

52. *Investigations of Dirigible Disasters: Hearings before a Joint Committee to Investigate Dirigible Disasters*, 73rd cong. 1st sess. (Washington, DC: USGPO, 1933), 203–4, 696.

53. For a more detailed discussion of the work of the Science Advisory Board in this matter, see Harper, "Boundaries of Research," 57–64; David Hart, *Forged Consensus: Science Technology and Economic Policy in the United States, 1921–1953* (Princeton: Princeton University Press, 1998): 72–75; Robert Kargon and Elizabeth Hodes, "Karl Compton, Isaiah Bowman, and the Politics of Science in the

Great Depression," *Isis* 76 (1985): 300–318; Carroll W. Pursell, "The Anatomy of a Failure: The Science Advisory Board 1933–35," *Proceedings of the American Philosophical Society* 109 (1965): 342–351.

54. Isaiah Bowman et al., "The Work of the Weather Bureau," *Science* 78 (22 Dec. 1933): 582–585, 604–607, quote on 584.

55. Ibid., 606.

56. Whitnah, *History of the Weather Bureau*, 160.

57. Ibid., 161.

58. George R. Stewart, *Storm: A Novel* (Lincoln, Nebraska: University of Nebraska Press, 1983 [1941]), 232. While *Storm* is fiction, Stewart was an eyewitness. According to one scholar, Stewart "secured introduction to the staff of the Weather Bureau in San Francisco, visited them during storms, and learned to draw his own weather maps" (John Caldwell, *George R. Stewart*, vol. 46, *Boise State University Western Writers Series* (Boise, ID: Boise State University, 1981), 30.) Jack Thompson, a meteorologist in the San Francisco office during the 1930s, remembered Stewart as a "familiar visitor around the office for a while" (Thompson, "Local Weather Bureau Office," 1252).

59. Namias, "The Early Influence of the Bergen School on Synoptic Meteorology in the United States," 492.

60. Biographers have occasionally noted that Reichelderfer gave up a promising navy career in which he was second in command of a battleship. While he was second in command, his ship, the USS *Utah*, had been decommissioned as a battleship following the London Naval Treaty of 1925. The big guns were removed, radio steering gear and extra armor were added. The ship became a mobile target. By 1938, when Reichelderfer was serving aboard, *Utah* had been fitted with various small guns to serve as an anti-aircraft training vessel. *Utah* retained her target capability, however. In August 1937, she was pummeled by 40 practice bombs dropped by Army Air Corps B-17s, and in September 1938 she was attacked by an early air-to-surface guided missile. Japanese pilots also mistook *Utah* for a battleship, and sank her at Pearl Harbor (Myron J. Jr. Smith, *Battleships and Battle-Cruisers, 1884–1984* (New York: Garland Publishing, Inc., 1985), 594, 596, 618.)

61. *Statutes at Large* 1014 (1938), cited in Whitnah, *History of the Weather Bureau*: 161.

62. Reichelderfer, "The How and Why of Weather Knowledge," 138–139.

63. Carl-Gustaf Rossby, "The Scientific Basis of Modern Meteorology," in *Climate and Man: Yearbook of Agriculture 1941*, ed. F. W. Reichelderfer (Washington, D.C.: Department of Agriculture, Government Printing Office, 1941), 599.

64. Ibid., 600.

65. The primary synthesis on 20th century American military meteorology is John F. Fuller, *Thor's Legions: Weather Support to the U.S. Air Force and Army, 1937–1987* (Boston: American Meteorological Society, 1990).

66. Fuller, *Thor's Legions*, 30.

67. Ibid., 51.

68. C. G. Rossby, "Preliminary Report on the Activities of the University Meteorological Committee," January 24, 1943. Box 3, University Meteorology Committee Papers (MC 511), Institute Archives and Special Collections, MIT Libraries, Cambridge, Massachusetts (Henceforth, UMC).

69. The lengthiest secondary source for the history of the wartime weather training is Raymond Walters, *Weather Training in the AAF: 1937–1945, US Air Force Historical Study No. 56* (USAF Historical Division, Air University, 1952). A livelier account is Diane Rabson, "It Happened Here: The Invisible Ally," *Staff Notes Monthly (University Corporation for Atmospheric Research)* October 1998. Available online at http://ucar.edu/communications/staffnotes/9810/here.html

70. Radio script, UMC, box 3. Hand notation says the program was given 4 Feb. 1943, on the Mutual Network, Station WGN. Diane Rabson discusses this in "The Invisible Ally."

71. "Announcement of Special Army-Sponsored Meteorology Training Programs," Undated, probably 1943. UMC box 3. For more on the preparation programs, see: R. E. Rowland, *The Premeteorology Program of the Army Air Force, 1942–1945* (http://www.rerowland.com/premet.html, retrieved 28 April 2004).

72. University of Chicago curriculum, 1943. This forty-nine page description of the curriculum of the Chicago A level school gives a detailed, week by week overview of what students were taught. "Syllabus of Courses Comprising the Training Program of Meteorologists for the United States Army Air Corps and Aerologists for the United States Navy," 32–56, Box 1, Folder 46 "Sverdrup, Misc.—Meteorology Programs 1943–1945." Scripps Institution of Oceanography Archives, San Diego, California.

73. The students picked up on this tension as well. "Throughout our life at NYU an indecisive battle royal raged between the academic and the military for our poor GI souls," wrote students in one yearbook (*Synopsis: Class 2-A-44, 30th AAF Training Detachment* [New York University, June 1944.], 3.)

74. Ibid., 8

75. This list is representative of the books used in other programs. See Chicago curriculum (n. 72) for another example. Chicago used Helmut Landsberg's *Physical Climatology* instead of Kendrew, and used the manuscript of Byers' *General Meteorology* instead of his earlier *Synoptic and Aeronautical Meteorology*.

76. The development of this one hundred-page pamphlet is described in Bernhard Haurwitz, "Meteorology in the 20th Century: A Participant's View (Part III)," *Bulletin of the American Meteorological Society* 66 (1985): 501.

77. These include Jörgen Holmboe, George E. Forsythe, and William Gustin, *Dynamic Meteorology* (New York: John Wiley and Sons, Inc., 1945); Sverre Petterssen, *Weather Analysis and Forecasting: A Textbook on Synoptic Meteorology* (New York: McGraw-Hill, 1940); Victor P. Starr, *Basic Principles of Weather*

Forecasting, ed. Carey Croneis, *Harper's Geoscience Series* (New York: Harper & Brothers, 1942); and Hurd C. Willett, *Descriptive Meteorology* (New York: Academic Press, Inc., 1944).

78. Starr, *Basic Principles of Weather Forecasting*, 2.

79. Ibid., 1.

80. Carl-Gustaf Rossby to Col. H.H. Bassett, March 13, 1943. UMC box 1. Original emphasis.

81. Fuller, *Thor's Legions*, 64, 191–192.

82. *Synopsis: Class 2-A-44, 30th AAF Training Detachment.*

83. The fullest source on African-American weathermen is Gerald A. White, Jr., "Tuskegee Meteorologists in World War II," article in preparation, February 2006. My thanks to him for sending it to me prior to publication. See also Fuller, *Thor's Legions*, 229–230. Charles Anderson went on to earn a Ph.D. after the war, the first African American meteorologist to do so.

84. The best studies of women in wartime meteorology are J. M. Lewis, "Waves Forecasters in World War II (with a Brief Survey of Other Women Meteorologists in World War II)," *Bulletin of the American Meteorological Society* 76 (1995): 2187–2202 and Kathleen Broome Williams, *Improbable Warriors: Women Scientists and the US Navy During World War II* (Annapolis, Maryland: Naval Institute Press, 2001). Three women had notable careers in post-war meteorological research. Joanne Simpson became the first American woman to earn a Ph.D. and has had a lengthy career in hurricane and tropical weather research. Florence Van Straten headed the Navy Weather Service's technical requirements section from 1948 to 1962. Dorothy Bradbury obtained a master's degree from Chicago in 1951, where she worked as a research scientist until retiring in 1974. Joanne Simpson's Oral History, part of the American Meteorological Society's Tape Recorded Interview Project, is available online at: http://www.ucar.edu/archives/publications/simpson-joanne%20interview.pdf. Some further context can be found in Fuller, *Thor's Legions*, 227–228; and Kaye O'Brien and Gary K. Grice, eds., *Women in the Weather Bureau During World War II* (Washington, D.C.: National Weather Service, 1991).

85. A useful source for thinking about the uses of yearbooks is Mariaelena Bartesaghi, "Reconstructing the High School Experience: The Role of Yearbooks in the Social Construction of Memory" (Master's Thesis, University of Pennsylvania, 1992). I thank her for an interesting conversation as well.

86. Colver R. Briggs, Arden Lanham, and Bruce Heater, eds. *Class 5: Meteorology Cadets* (University of California, 1943). UMC, box 5, "UCLA".

87. *Synopsis*: 17.

88. *Synopsis*: 25.

89. For examples, see SIO Office of the Director (Sverdrup), 82–56, Box 1, Folder 46 "Sverdrup, Misc.—Meteorology Programs 1943–1944," Scripps Institute of Oceanography Archives.

90. Here the Council was declaring that it had decided against Rossby's earlier thought that the Society ought to be broken apart.

91. For instance, see Spence, *The Rainmakers* or Katharine Anderson, *Predicting the Weather: Victorians and the Science of Meteorology* (Chicago: University of Chicago Press, 2005).

92. Horace R. Byers, "Recollections of the War Years," *Bulletin of the American Meteorological Society* 51 (1970): 216.

93. David D. Houghton, "Meteorology Education in the United States after 1945," in *Historical Essays on Meteorology, 1919–1995*, ed. James R. Fleming (Boston: American Meteorological Society, 1996), 541–553.

94. Clark A. Miller, "Scientific Internationalism in American Foreign Policy: The Case of Meteorology, 1947–1958," in *Changing the Atmosphere: Expert Knowledge and Environmental Governance*, ed. Clark A. Miller and Paul N. Edwards (Cambridge: MIT Press, 2001), 167–218.

95. Horace R. Byers, "Chairman's Prefatory Remarks," prior to the Symposium on Modern Concepts in Meteorology, April 27th, 1955. Proceedings of the National Academy of Sciences, v. 41, n. 11 (November 15th, 1955): 797.

96. Edward N. Lorenz, "The Evolution of Dynamic Meteorology," in *Historical Essays on Meteorology, 1919–1995*, ed. James R. Fleming (Boston: American Meteorological Society, 1996), 3–19.

CHAPTER 7

The Struggle over Airways in the Americas, 1919–1945

Atmospheric Science, Aviation Technology, and Neocolonialism

GREGORY T. CUSHMAN

BEYOND COLONIAL SCIENCE

Why are Latin America and Canada so marginal to our understanding of the history of science and technology? From the beginning, Latin America's rulers have had technocratic tendencies as strong as those of any world region. In 1822, less than a year after sending his botanist vice-president to Europe to import a commission of scientific advisors, Simón Bolívar repeated Humboldt's famous ascent of Chimborazo in order to gain a bird's-eye view of the vast territory he had liberated from Spanish rule.[1] We have ignored Latin America's scientific contributions at our own peril. The inhabitants of Galveston, Texas, for example, might have appreciated knowing that Jesuit meteorologists in U.S.-occupied Cuba had warned of a hurricane heading in their direction in September 1900.[2] This chapter will examine a 25-year period when the skies of the Western Hemisphere, from the Arctic Ocean to Cape Horn, became a major concern to the industrial world's most influential aviators and atmospheric scientists.

But the main reason for this marginality stems not from the poverty of science and technology in these regions, but instead from the *poverty of*

175

concepts we employ to understand the geopolitical ramifications of environmental knowledge and technological development. Thanks to a flurry of recent scholarship, it is no longer acceptable to dismiss the contribution of colonial science and technology to modern history.[3] But except for "white settler colonies" like Australia, South Africa, the U.S., and Canada, historians have given little attention to science and technology in postcolonial contexts.[4] This poverty of research makes it difficult to answer postcolonial critics who claim that Western science and technology are, in essence, tools of empire for subjugating indigenous peoples and knowledge systems to the will of the "laboratory state."[5] This chapter will introduce the concept of *neocolonialism* to this discussion.

As an analytical concept, neocolonialism seeks to explain the lasting external dependence of postcolonial states and peoples, as well as the long historical persistence of global power structures after formal decolonization. Neocolonialism (or "informal empire," as it is sometimes known), also represents a particular historical era in the relationship between the Great Powers and the rest of the world. Neocolonialism has been called "the American stage of colonialism" in recognition of the marked tendency of U.S. international relations to take neocolonial forms, particularly in the Western Hemisphere during the first half of the twentieth century.[6] Latin American historians use this analytical concept with great frequency to make sense of challenges to national sovereignty faced by the former colonies of Spain, Portugal, and France since Haiti's independence in 1804, and they often refer to the period 1870–1930 as Latin America's neocolonial epoch.[7] In contrast, Canadian historians seldom use this term, though it may be useful to consider "Americanization," a long-standing Canadian concern, as a variety of neocolonialism.[8] These considerations all point to the importance of Latin America, Canada, and their relationship to the United States for any general understanding of postcolonial science and technology.

Neocolonialism has several specific historical tendencies. Neocolonial relationships often lacked centralized supervision, much less control, by state-level powers. Foreign business and financial interests were often paramount in these relationships, though the exchange of education, expertise, and technological aid could also play an important role. Typically, local elites were critical to the establishment and maintenance of neocolonial relationships and often profited richly from them. The actors involved often did not consciously intend to encourage inequality. (Of course, much the same can be said for the haphazard, pragmatic way in which earlier colonial relationships developed.) However, few neocolonial theorists have recognized the ability of locals to play one neocolonial power against another. This opportunity makes neocolonialism distinct from formal colonial rule, and it potentially provided local actors with substantial "room for maneuver" to avoid external coercion—as the following case study will show.[9]

From the end of World War I to the end of World War II, the airways of Latin America and Canada served as a major battleground for neocolonial ambitions. U.S. aviation companies won this struggle, largely at the expense of German settler enclaves in the Americas. (In the process, "airminded" Germans refocused their colonialist ambitions away from South America and scientific internationalism toward the lands and airways of Central and Eastern Europe—and another World War.[10]) Atmospheric science attained enormous strategic value in this geopolitical context. A small group of Scandinavian scientists and their disciples took decisive advantage of this situation. The "Bergen school" of meteorology used this conflict to conquer an entire hemisphere—the Western hemisphere—for a particular approach to understanding the atmosphere based on polar front theory and air mass analysis. Their triumph led directly to the establishment of new centers of action for atmospheric research in North America, as well as the rapid expansion of scientific capabilities in Latin America and the Caribbean, even as it marginalized some promising research programs in tropical meteorology. To a lesser extent, this study will explore the following themes: (1) the geopolitical ramifications of the migration of European scientists and the formation of new settler enclaves and research schools in the Americas, (2) the influence of less spectacular (and often overlooked) aviation technologies on atmospheric science, (3) the potential for science and technology to improve the social status of individuals, and (4) the contingent relationships linking scientific to technological development.

In stark contrast to the determinism of most studies of neocolonial engagements, this history will emphasize the variety of outcomes that arose from this neocolonial struggle over atmospheric science and aviation technology in the Americas. German scientists used neocolonial expeditions to the Americas to develop the world's leading center for maritime meteorology. From the point of view of many practicing meteorologists in the United States, their country was forcibly colonized by an unwelcome, "Norwegian" science between the wars. Both Norwegian and Canadian meteorologists, on the other hand, used polar front theory to assert their scientific autonomy. As a result of World War II, Brazil acquired a vast network of modern airports and meteorological observatories; Argentina became a haven for talented refugee meteorologists from Germany; while parts of Ecuador were forcibly occupied by the United States to enforce its power over the airways of the equatorial Pacific. Adherents of the Bergen school working in colonial Puerto Rico, meanwhile, discovered that polar front theory had little to teach tropical scientists and founded a new discipline: modern tropical meteorology. Depending on context, science and technology served as potent tools for external domination as well as for postcolonial liberation.

In the eyes of most historians, the World Wars and Great Depression precipitated crisis, if not collapse, for the Age of Empire. This was not true for the atmospheric realm. The symbiosis between science, technology, and external domination flourished during the Age of Aviation. Meteorology, in many ways, became a neocolonial science between the wars. This only becomes clear when its historical development is examined from a global perspective.[11]

"TRANSPORTATION IS CIVILIZATION"

Unlike North Americans, who gave birth to the Wright Brothers, Latin Americans are often considered laggards when it comes to embracing modern science and technology. Yet Brazilian aviation enthusiasts will happily inform you that their compatriots accomplished an impressive list of "firsts" using craft of their own design, including the first launch of a paper airship before the court of João I in Lisbon in 1709.[12] But most innovations in aviation technology and atmospheric science have diffused from Europe and the United States to the rest of the Americas.[13] These transfers took place with extraordinary rapidity and by a remarkable variety of avenues. News of the Montgolfier brothers' first balloon flights in France reached Mexico within months, helped along by science-oriented newspapers signifying Spanish America's Enlightenment. The *Gaceta de Méjico* reported the flight of hot-air balloons—and an occasional aeronaut willing to risk immolation—in five Mexican cities in 1785. That same year, in Concepción, Chile, the French explorer La Pérouse launched the first known balloon in South America as a display of French scientific prowess and imperial ambition in the Pacific. During the 1830s and 1840s, self-proclaimed "professors of physics" launched balloons in major cities all over the hemisphere. Some American aeronauts used balloons to make scientific observations of the upper atmosphere from an early date. Adolfe Theodore, a French-born mulatto, tested the effects of altitude on the behavior of birds and the flavor of volatile spirits during an early hydrogen balloon flight above Havana in 1830. (He determined that

Cuban rum tastes just as good when one approaches the heavenly realm.) Latin Americans also embraced airplanes with similar rapidity and flair. In November 1911, two years after a rich playboy from a powerful Mexican political family made the inaugural airplane flight in Latin America, Mexico's President Francisco Madero became the first head-of-state in the world to fly in an airplane.[14] Scientific men like Madero, a Berkeley-trained agronomist, would have enthusiastically embraced the motto "Transportation Is Civilization." Latin America's postcolonial rulers have long lusted after the power, prestige, and progress promised by new technologies, especially railroads. The transportation revolution they helped engineer contributed mightily to the emergence of a neocolonial order spanning the region after 1870.[15]

But one crucial innovation in aviation technology, particularly where meteorology is concerned, diffused from the southern periphery. In 1883, after spending a decade extending the scientific reach of the British Empire in Oceania, Lawrence Hargrave retired from the Sydney Observatory to dedicate himself to human flight. In stark contrast to the Wright Brothers (whom he influenced), Hargrave was philosophically opposed to proprietary research and enthusiastically shared his designs internationally. In 1893, he invented the box kite and reported this to the World's Columbian Exposition in Chicago. U.S. kite designers immediately began corresponding with Hargrave and initiated meteorological observations using kites at Blue Hill Observatory near Boston in 1894. The Hargrave kite quickly emerged as the preferred instrument platform for systematic observation of the upper atmosphere because of its simplicity, durability, stability, superior lifting prowess—and lack of patent protection. Their use soon spread to exotic locations, following the tentacles of empire: to India and the tropical Atlantic in 1905, Samoa and Spitsbergen in 1906, Lapland and Siberia in 1907, East Africa and Java in 1908. Kites remained standard equipment for first-order observatories in the United States until 1933.[16]

The Second Latin American Scientific Congress in March 1901 provided direct stimulus for the initiation of upper-air meteorology in South America. Delegates from eleven Latin American republics celebrated the opening of a new meteorological observatory at Montevideo's Alameda Park. Later that year, the observatory's first director, Luis Morandi, began making observations using Hargrave's kite designs, albeit with disappointing results. Meteorologists in Rio de Janeiro and Buenos Aires followed suit in 1905 and 1910. In 1909, Morandi also became the first Latin American to experiment with rubber weather balloons—one year before the Canadian

Meteorological Service initiated kite and sounding balloon observations. Like kites, sounding balloons carry recording instruments, but they are often untethered and can reach far greater altitudes. Morandi quickly abandoned their use when it proved difficult to recover their expensive, imported instruments in the sparsely inhabited pampas. He had much better success with pilot balloons, a technique to measure the speed and direction of upper-level winds by tracking expendable balloons with a theodolite (Figure 1). By 1912, Morandi had made a total of 31 launches using German balloons when Montevideo's only hydrogen plant closed down and forced him to discontinue this line of work—a symbol of the material barriers that have long faced Latin Americans working on the cutting edge.[17]

"THE LAST FREE CONTINENT"

The end of World War I brought the world's first aviation boom to a crashing halt. Unemployed military pilots and mechanics scattered to the ends of the earth to start dozens of airlines. Great Britain, France, Italy, the Netherlands—even Switzerland—quickly located outlets for their accumulated expertise, industrial capacity, and surplus aircraft: either by providing rapid transportation and "air control" for their overseas empires, or by sending postwar air missions to Latin America explicitly aimed at diluting the influence German military missions had acquired before World War I.[18] After purchasing 85 French warplanes in 1921, the Brazilian government made a half-hearted attempt to centralize its meteorological service and produce aerological forecasts, starting with a kite observatory strategically placed close to Brazil's southern border with Argentina and Uruguay. Argentina invited military air missions and instituted systematic upper-air surveys of its own in order to keep up with its powerful neighbor.[19]

The Versailles Treaty closed off most of these options to Germans. Even before the end of the war, various economic interests in Germany and immigrant groups abroad initiated a conscious reengagement with Latin America, "the Last Free Continent" open to large-scale German investment and influence. The new international relationships they formed exhibited classic features of neocolonialism: the primacy of economic issues, the centrality of business interests and settler enclaves, the unbalanced exchange of primary materials for high-tech industrial goods, and the lack of formal, centralized control. Most significantly for the history of science and technology, *experts* played a critical role in fostering these relationships, usually by coordinating local and external interests.[20]

In December 1919, a group of Colombian and *Auslandsdeutschen* investors formed the Sociedad Colombo-Alemana de Transportes Aéreos. SCADTA made its first flight from the coast into the interior on 19 October 1920 after the importation of two Junkers F-13s—an all-metal, four-passenger monoplane explicitly designed for German export—and an inventive team of pilots and mechanics from Germany. SCADTA's efforts received an enormous boost with the arrival of Peter Paul von Bauer, an ambitious, wealthy Austrian scientist with a newly minted Ph.D. and prewar map-making experience in the Colombian Amazon. Unlike the meteorologists described in Deborah Coen's article in this volume, Bauer looked abroad for opportunities after the dissolution of the Austro-Hungarian Empire. He purchased a controlling share of SCADTA and two more F-13s altered to specifications suggested by SCADTA technicians to make them suitable for Colombian conditions. Additional cross-Atlantic financing and technical support from Junkers Flugzeugwerke AG and Deutschen Petroleum AG enabled the survival of this fledgling company. A lucrative air mail subsidy, loans for route expansion, and vociferous support from the Colombian government (over repeated French and U.S. protests) turned SCADTA into the most successful early airline in the Western Hemisphere.[21] Junkers tried to duplicate this triumph with aviation missions to the Southern Cone, Cuba, Mexico, Venezuela, Brazil, and Bolivia. SCADTA and Lloyd Aéreo Boliviano (LAB) together purchased 34 F-13s, one-tenth of the total production run of this revolutionary aircraft. This entailed the continued use of Junkers parts, loans, and technical support, which encouraged, in turn, the purchase of later models. German-supported airlines utterly dominated commercial air travel in the region during the 1920s (Table 1), converting the German city of Dessau (the headquarters of Junkers and the Bauhaus movement) into the metropole for Latin American aviation technology. These overseas endeavors not only enabled Junkers to survive ruthless postwar competition and economic upheavals in Europe, they convinced the Weimar government to provide official support to Junkers and Latin American aviation, most notably through the incorporation of Deutsche Luft Hansa (1926). For Rinke, this was a definitive case of the formation of German foreign policy "from below."[22]

Science played an important supporting role in these endeavors by strengthening the perception that Germany's technological advancement

TABLE 1   Ownership of Major Latin American Passenger Airlines, 1927–1940

	route miles 1927	miles flown 1927	route miles 1934	miles flown 1934	route miles 1940	miles flown 1940
Pan American Airways Corp.[a]	110	15,000	25,009	6,625,000	41,619	13,801,000
Other U.S.[b]	412	n.d.	4,608	1,652,000	6,156	2,311,000
German[c]	2,574	399,000	5,332	1,338,000	12,890	2,291,000
Latin American (German equipped)[d]	0	0	2,792	576,000	5,167	1,273,000
Other Latin American[e]	0	0	8,700	1,242,000	19,809	3,809,000
British Commonwealth[f]	0	0	1,700	600,000	5,000	2,300,000
Dutch[g]	0	0	0	0	1,982	599,000
Italian[h]	0	0	0	0	6,125	590,000
French[i]	2,200	35,000	2,900	327,000	0	0
Total	4,884	449,000	51,549	12,472,000	98,748	26,974,000

Sources: William A. M. Burden, *The Struggle for Airways in Latin America* (New York: Council on Foreign Relations, 1943), tables 6, 15, 53, maps 4, 6–7; R. E. G. Davies, *Airlines of Latin America since 1919* (Washington, DC: Smithsonian University Press, 1984), app. 3.

[a] Pan American Airways, Pan American-Grace Airways, and local subsidiaries: Mexicana, Cubana, Aerovías de Guatemala, Panama, SCADTA/Avianca (after 1931, Colombia), UMCA (Colombia), Panair do Brasil.

[b] Local airways owned by resident U.S. citizens: Aeronaves de México, Cía. Aeronáutica del Sur (Mex.), LAMSA (Mex.), Pacífico (Mex.), West Indian Aerial Express, Caribbean-Atlantic, Faucett (Peru), Condor Peruana.

[c] Includes German-owned airlines: Syndicato Condor (Brazil), SEDTA (Ecuador), Lufthansa Sucursal (Peru), Deutsche Lufthansa, and locally owned airlines with significant German financial interest, staff, and equipment: SCADTA (until 1931, Colombia), VARIG (Brazil), LAB (Bolivia).

[d] Locally owned private and government airlines flying German planes: VASP (Brazil), Aeroposta Argentina, LASO (Argentina), LAN-Chile, CAUSA (Uruguay).

[e] Locally owned private and government airlines: includes 5 Mexican, 2 Brazilian, 1 Panamanian, 1 Venezuelan, 1 Peruvian, 1 Uruguayan, and 1 Argentine airlines.

[f] British West Indian Airways and TACA (Central America), an airline owned by a New Zealand-born Canadian WWI veteran.

[g] Royal Dutch Air Lines (KLM).

[h] LATI and Corporación (Argentina).

[i] Lignes Latécoère and Air France.

stemmed from its scientific prowess. Bauer organized a Scientific Section as one of his first acts as SCADTA director. He surreptitiously imported a complete set of aerial photographic equipment from Germany and immediately set his staff to mapping local air routes. In 1923, SCADTA offered its services to the Swiss scientific mission surveying the Colombia-Venezuela border. This generated enormous goodwill among nationalist Colombian officials who later granted lucrative concessions to map the Panamanian border and vast Colombian Amazon. The Venezuelan state, in turn, bought its first civilian airplane, a Junkers W-34, to do its own cartographic work on the Colombian border.[23]

Meanwhile, meteorologists based at the Deutsche Seewarte (German Marine Observatory) in Hamburg coordinated a vast research project to aid German trade with the Americas. Two giants of German environmental science, Wladimir Köppen and Alfred Wegener, hatched a plan during the closing days of the Great War to equip German merchant vessels for observing upper-level weather conditions in the shipping lanes of the Atlantic. This served Köppen's long-standing goal to develop a three-dimensional typology of the world's climatological zones. Wegener directed the first Deutsche Seewarte meteorological expedition to the Gulf of Mexico in spring 1922—just in time for Junkers's aviation missions to Cuba and Mexico.[24]

José Carlos Millás, the energetic hurricane expert who directed the Cuban National Observatory, was impressed when Wegener made the first pilot balloon soundings ever performed in Havana. This provided a far more direct method for keeping track of upper-air currents than the nephoscopes Cuban meteorologists had been using since the early 1880s to follow cloud movements. In 1925, Millás approached the U.S. Weather Bureau, "the best Meteorological Service in the world," for help. The head of the U.S. Weather Bureau provided apparatus and several U.S. publications on aerological meteorology. On 26 January 1926, after receiving further practical advice from British meteorologists in Jamaica, the Havana observatory initiated daily pilot balloon observations, to great public acclaim, and began exchanging them with Kingston during the subsequent hurricane season (Figure 1). Following the lead set by his predecessors, Millás immediately began using this data to hone his theories regarding hurricane structure and movement.[25]

In 1925, SCADTA finally received grudging permission from U.S. officials to fly across the Panama Canal Zone, the centerpiece of the United States' neocolonial empire in the Caribbean. With financial support from the Colombian government and the Condor Syndikat, a new German company founded to promote Latin American aviation, Peter Paul von Bauer began

FIGURE 1   Upper-air observations at the Cuban National Observatory, Havana 1926. Top: Filling. A rubber pilot balloon filled with hydrogen must be carefully weighed to assure an accurate ascent rate. This apparatus was donated by the U.S. Weather Bureau. Middle: Tracking. Theodolite used to track an ascending pilot balloon. Note retractable, "hurricane-proof" cover. Bottom: Analysis. Cuba's first team of aerological observers, José Santiago (left) and Gustavo Castillo (right), plotting a pilot balloon trajectory for deriving upper-level wind speed and direction. In *Boletín del Observatorio Nacional* (Havana) 22, no. 1 (1926).

implementing his long-standing plan to establish the first airline connecting South America and the United States. The Deutsche Seewarte again provided timely scientific support. During its fifth expedition to Latin America, German meteorologists established contact with SCADTA and began arranging to use its planes to survey atmospheric conditions above the southern Caribbean. A sixth expedition arrived in Colombia in June 1925 and stayed for five weeks. In the course of seven survey flights, Walter Georgii and his assistant Heinrich Seilkopf made the most extensive aerological investigation of the tropical atmosphere that had ever been made in the Americas. In addition, they provided a crash meteorological course to SCADTA's pilots. After leaving Colombia, they made a double transect by ship along the Central American coast all the way to Guatemala and made a valuable discovery: a persistent line where upper-level winds abruptly shift direction during the summer in the western Caribbean: a "cyclonic shear zone." These observations paved the way for SCADTA's celebrated flight from Colombia, through Central America and Cuba, to Florida in August–September 1925.[26]

Together, these 1925 expeditions exercised a major influence on the relationship between aviation, atmospheric science, and neocolonialism in the Western Hemisphere. Despite tentative support from U.S. Department of Commerce officials, Bauer failed to obtain landing rights in the United States for a regular trans-Caribbean service. Following a well-established pattern, the U.S. War Department portrayed SCADTA's Germanic ties as a violation of the Monroe Doctrine and a major threat to the Panama Canal; they refused to tolerate any "Pan-American" airline in the Caribbean that was not under total U.S. control.[27] German aviation interests and the Deutsche Seewarte gave in to this reality and shifted their focus away from the Caribbean. From 1922 to 1930, the Deutsche Seewarte completed ten meteorological expeditions to Latin America, five to West Africa, and several more to the North Atlantic (Figure 2), and played an important role in the celebrated 1925–1927 *Meteor* expedition to the South Atlantic. (Wegener died on an expedition to the Danish colony Greenland.) Their work paved the way for new German airlines on the eastern coast of South America and for the first trans-Atlantic flights by catapult mail planes and the *Graf Zeppelin*.[28] Following their lead, Argentine aviation boosters with colonial aspirations in Antarctica planned a meteorological expedition to the South Pole for the end of 1926. But the German hydroplane slated to fly meteorological instruments above the southern continent crashed before it arrived in Argentina and quashed these ambitions.[29]

186

FIGURE 2   Meteorological Research Voyages by the German Oceanographic Observatory, 1922–1928. Gulf of Mexico (Azores, Cuba, Mexico, U.S., 3 trips, 1922–23), Atlantic South America (Canary Is., Brazil, Uruguay, Argentina, 3 trips, 1924–25, 1927–28), Caribbean Sea (Trinidad, Venezuela, Curaçao, Colombia, Panama, Costa Rica, Nicaragua, Honduras. Guatemala, Haiti, 3 trips, 1924–25, 1927), West Africa (Canary Is., Liberia, Sierra Leone, Portuguese Guinea, French West Africa, Gambia, 2 trips, 1928), New York (1 trip, 1928), Iceland/Greenland (3 trips, 1926–28). Not shown: *Meteor* expedition to the equatorial and southern Atlantic (1925–27). Photo: 1931, courtesy of Bundesamt für Seeschiffahrt und Hydrographie, Hamburg.

Germany's interest in the airways of the Western Hemisphere laid foundations for a major change in German meteorological practice, largely through the efforts of Heinrich Seilkopf and his disciples. After completing his habilitation thesis on the Deutsche Seewarte's first six meteorological voyages to the Americas, Seilkopf became a standard fixture on important

German flights, including two inaugural voyages by the *Graf Zeppelin* to Brazil in 1931. Meanwhile, he oversaw the marked expansion of the Deutsche Seewarte's maritime meteorological staff and network of ship-borne aerological observers in the Atlantic Basin. Most significantly, a group of young meteorologists under Seilkopf's supervision developed a powerful new forecasting technique keyed on changes in the topography of barometric pressure in the upper atmosphere revealed by these maritime surveys. In 1940, these "isobaric" techniques supplanted "Norwegian" air-mass analysis as the preferred tool used by the Reichswetterdienst to direct the Luftwaffe's long-range bombing campaigns.[30] Thus, neocolonial meteorology not only increased German prestige—and deadly force—abroad, it influenced the basic content of the "exact sciences" back home.[31] By appropriating the weather in the Atlantic Basin, the Deutsche Seewarte transformed itself into the world's unrivaled center for maritime meteorology.

As with so many aspects of international affairs, the United States was late to join this struggle, but it made a decisive impact once it did. Rather than sending atmospheric scientists as its advance guard, the U.S. War Department sent military fliers. In response to SCADTA's 1925 trip, on 22 December 1926, five Loening hydroplanes crossed the Rio Grande for a grand tour that touched down in every Latin American republic. This "Pan American Good Will Flight" sought to generate publicity, to survey potential commercial routes, and, internal documents reveal, to counter "alien" aerial activity in the region. The result was a public relations disaster. U.S. marines had directly intervened in the affairs of nine Latin American states since 1919, and locals clearly detected the "military character" of the flight. Army Air Force fliers met a frosty reception in many locations. To make matters worse, their planes broke down repeatedly, and two pilots died in Argentina after two Loenings collided in flight—hardly a strong showing of technological prowess.[32]

This perception changed overnight in 1927 when Charles Lindbergh, "the Columbus of the Air," piloted *The Spirit of St. Louis* non-stop from New York to Paris. Six months after Lindbergh's historic flight, he made a complete circuit around the Caribbean timed to inaugurate the Sixth Pan American Conference in Havana and Pan American Airways' new passenger service from Key West to Havana.[33] Through a lucrative $2 per mile subsidy, the U.S. Postal Service ensured that a single U.S.-controlled airline acquired a virtual monopoly over North-South air service in the Americas. Pan American Airways used this enormous political and financial leverage to purchase the latest U.S. aircraft and to buy up a long list of regional airlines. At the nadir of the Great Depression, Peter Paul von Bauer reluctantly sold 84 percent of SCADTA's shares to Pan American Airways. This brought an end to SCADTA's special relationship with Junkers Flugzeugwerke and turned SCADTA into a consumer of second-hand U.S. aircraft. But Pan American shrewdly retained SCADTA's German staff and profited from their long experience, their contractual obligation to work at a much lower pay scale than U.S. employees, and their utility in obscuring this northern takeover from Colombian nationalists. As Table 1 shows, by 1934 Pan American Airways and its subsidiaries had became as large as the rest of the region's commercial airlines put together. Profits from Latin America, in turn, enabled Pan American to offset enormous operating losses accrued by its pioneering East-West transoceanic routes during the late 1930s.[34]

Science played a role in the meteoric rise of Pan American Airways. It immediately stood to benefit from the new pilot balloon observatory in Havana, and its partner in flight along the Pacific Coast, W. R. Grace & Co., already had experience reporting maritime observations from Grace Lines ships meant to keep track of El Niño.[35] But instead of depending on existing scientific institutions in the region, Pan American Airways preferred to organize its own, private meteorological network. By 1933, its subsidiary Mexicana had come to operate twelve pilot balloon observatories and five radio stations issuing weather reports at Mexican airports. Pan American Airways also used science to generate publicity. In 1929, Lindbergh made a five-day photographic survey of lowland Mesoamerica in search of remote Mayan ruins in a Pan American hydroplane. Most significantly, Pan American meteorologists aggressively applied "standard high-latitude methods of frontal analysis" to tropical weather, particularly in the zone of equatorial wind convergence, a phenomenon christened by Norwegian meteorologists as the "inter-tropical front."[36]

Latin American politicians, businessmen, and soldiers made their own mark on this neocolonial struggle. Local capitalists founded dozens of small, regional airlines, and even rabid nationalists had good reasons for embracing aviation. Better transportation promised increased trade and greater local incomes (even if a tiny elite got the lion's share). Like railroads for previous generations, airplanes represented the cutting edge of modern technology and potent symbols of progress. Thus, Brazil's nationalist president Getúlio Vargas saw fit to spend US$1 million to build a terminal outside Rio to receive German dirigibles in 1936, only to close it a year later in the wake of the *Hindenburg* disaster. Vargas also used a German hydroplane to fly around Brazil campaigning for political support, an eventual key to his transformation into Brazil's definitive populist politician. Economic nationalists made sure these flying enclaves were not isolated from their host societies. Mexico, Brazil, and Chile led a trend requiring foreign airlines to employ local ground crews and administrators, train citizen pilots and mechanics, and maintain a level of national ownership. Even more than railroads, airplanes held enormous strategic importance for Latin America's rulers, since they provided a rapid, effective means to integrate national territory under a single, centralized power. (In the eyes of some critics, this enabled a potent form of *internal colonialism* by postcolonial governments.) To this end, the Peruvian, Chilean, and Brazilian militaries established their own airlines to serve regions either too remote or too important to leave to private companies. Argentina and Venezuela, meanwhile, took advantage of the Great Depression by taking control of routes and equipment orphaned by the bankruptcy of another contributor to this struggle, the French firm Aéropostale.[37]

Meanwhile, the growth of Latin American aviation indirectly exacerbated the gulf in technological innovation and industrial productivity separating the United States from its neighbors. During the mid-1930s, vigorous competition between U.S. airlines for the fastest, most efficient planes propelled a revolution in aerodynamic design. This revolution was led by the famous DC-series of aircraft, a product of close collaboration between hydrodynamic physicists at Cal Tech, airframe engineers at Douglas Aircraft in Santa Monica, California, and Transcontinental and Western Airlines (TWA). By the end of 1941, 260 of 322 scheduled airliners flying in the United States were DC-3s. To support this spending spree, U.S. airlines sold their obsolete aircraft abroad. This enabled Latin American airlines to expand their fleets while freeing the U.S. aviation industry to "push the envelope" technologically. In this way, "exports saved the air industry" in the United States during the Great Depression.[38]

German manufacturers could no longer compete on these terms. Instead, as an explicit part of National Socialist trade policy, Junkers offered slow but rugged JU-52 tri-motors and other craft at lower prices and much better repayment terms than its U.S. competitors. Deutsche Lufthansa's Brazilian subsidiary, Syndicato Condor, sweetened the deal by offering convenient servicing at shops in Rio and Buenos Aires. Meanwhile, Focke-Wulf arranged to open local assembly plants in Brazil and Argentina, thus supporting the beginnings of import-substituting industrialization. In Brazil, the most competitive market in Latin America, German aircraft maintained their numerical edge until World War II. Many expert observers, dazzled by U.S.

technological superiority, irrationally dismissed this inability to vanquish German aviation as a sign of Latin America's backwardness. Preference for German aircraft actually represented a shrewd strategy of postcolonial development on the cheap.[39]

These circumstances directly influenced atmospheric science. South American meteorologists, led by Argentina and Brazil, began organizing themselves internationally to keep up with the rapid expansion of commercial aviation during the 1930s. They looked for leadership to the International Meteorological Organization (IMO) based in Europe, as they established shortwave radiotelegraph networks to broadcast weather bulletins long distance and altered local observation practices to comply with IMO standards. This often meant looking to Germany for instruments and assistance. Pilot balloon observers were asked to coordinate their work with IMO Aerological Commission headquarters in Berlin. In December 1938, the Argentine Meteorological Service sent its master barometer (Wild-Fuess no. 100751) for calibration with official instruments in Berlin. Representatives at the Montevideo meteorological conference held the following February then resolved to calibrate most weather service barometers in South America with this Argentine instrument. To reinforce these international relationships, the Deutsche Seewarte permanently stationed German meteorologists at Syndicato Condor headquarters in Rio de Janeiro in 1935 and the Argentine Meteorological Service in 1938 and distributed its synoptic maps, free of charge, to all South American meteorological services.[40] Few South American meteorologists shared José Carlos Millás's lofty opinion of atmospheric science in the United States. Nevertheless, technology provided an entry point for U.S. influence: conference organizers asked the instrument division of Bendix Aviation, a long-time supplier to the U.S. Navy and Weather Bureau, to send an engineer to Montevideo to demonstrate radiosonde, the latest technique for sounding the upper atmosphere.[41] This was a portent of things to come.

THE COLONIZATION OF NORTH AMERICAN METEOROLOGY

Scandinavian science has long had imperial ambitions, particularly in the Arctic.[42] During World War I, Norwegian geophysicist Vilhelm Bjerknes originated the concept of an ever-undulating "polar front" demarcating the violent collision between cold and warm air masses. He hoped to use this concept to found a new, imperial form of meteorology based on hydrodynamic physics. Ideally, his home institution, the Bergen Geofysisk Institut—itself an expression of Norwegian scientific independence from Sweden—would become an "international central" charged with analyzing "the general weather conditions of the whole northern hemisphere."[43] But imperial rivalries prevented Vilhelm Bjerknes from accomplishing this hemispheric vision. Nevertheless, scientific apostles originally trained by Bjerknes at Bergen did succeed at colonizing an entire hemisphere for this "Norwegian school" of meteorology—the Western Hemisphere. This breakthrough represents one of the most dramatic geographical expansions of a research school in the history of science and underscores the importance of students to the success of a research school. Unfortunately, the literature on research schools has given little attention to their geographical propagation, due to its preoccupation with founding personalities and training institutions.[44]

Vilhelm Bjerknes's ambitions shifted decisively west toward North America during the 1920s. In 1924, he and his son Jacob visited U.S. Weather Bureau headquarters in Washington. Vilhelm put his son to work analyzing working charts in order to determine if polar front theory was applicable to North American conditions. Jacob tentatively concluded that it was, but needed more aerological data to be sure. In 1926, the Swedish-American Foundation paid another disciple, Carl-Gustaf Rossby, to spend a year at the U.S. Weather Bureau continuing this task. Rossby confidently attacked the most complicated U.S. weather maps and "furnished conclusive evidence that the polar front theory . . . enables us to explain phenomena which" otherwise "would hardly be understood."[45] Cash-strapped Weather Bureau officials were less than enthusiastic about Rossby's bold recommendation to overhaul the U.S. observation network to produce high-quality aerological data for air mass analysis. The Guggenheim Fund for the Promotion of Aeronautics, on the other hand, embraced this gospel and hired Rossby to develop an experimental "airway weather service" in California and establish a meteorology program at MIT. Ever the empire builder, Rossby imported Jacob Bjerknes, Sverre Pettersen, and Jörgen Holmboe to teach courses in the United States. Rossby and an occasional U.S. convert also made pilgrimages back to Norway to keep abreast of advances in Bergen-school techniques.[46]

Aviation thus provided a beachhead for the colonization of the U.S. meteorological profession by Scandinavian science. In 1935—the same year the German Weather Service autocratically imposed "Norwegian" air mass analysis on its forecasters—three of Rossby's American students were hired to inculcate these techniques at the U.S. Weather Bureau.[47] In 1938, one of the original U.S. converts to the Bergen school, U.S. Navy aerologist F. W. Reichelderfer, was appointed head of the U.S. Weather Bureau and hired Rossby as his assistant chief. As Roger Turner shows elsewhere in this volume, they used these positions to wage a successful campaign to transform U.S. meteorological culture along Nordic lines.[48] An essential element in the long-term success of a research school lies in its ability to conquer other scientific communities.

In July 1939, Jacob Bjerknes and his wife Hedvig left Norway for another extended tour of meteorological institutions in Canada and the United States.[49] It is highly significant that Jacob Bjerknes headed first to Canada. This vast country makes for a fascinating comparison with Latin America because in many ways it was still asserting its independence. After the carnage of World War I, autonomy from British foreign policy became the "linchpin" of Canadian national politics. Meanwhile, the Canadian economy was becoming far more dependent on the United States. Like Latin America, Canada experienced an economic boom during the 1920s based on the export of primary materials, particularly paper and metals produced on its northern frontier. Except for automobile assembly plants established by U.S. companies to evade imperial tariffs, Canadian industry fell into decline. Even during the height of the 1920s boom, lack of high-paying jobs, particularly in research and development, encouraged a brain drain across the U.S. border. Winnipeg was devastated when the Panama Canal, the most obvious symbol of U.S. neocolonialism in the Americas, diverted trade from Canada's celebrated transcontinental railways. Nevertheless, Presidents Coolidge and Hoover virtually ignored Canada, in stark contrast to the attention they lavished on Latin America. This was widely resented by Canadians, who also began to express concern about the impact of American film and radio on Canadian national culture. Even though criticism was muted, U.S. neocolonialism was a genuine threat to Canada.[50]

The same can be said regarding U.S. influence over Canadian meteorology and aviation. Once the British Empire's Magnetic Crusade exhausted itself in the 1850s, the development of meteorology in Canada became closely tied to the United States. Early Canadian observers reported to the Smithsonian, then to the U.S. Army Signal Corps' weather telegraphy network. For many years, Toronto's central observatory depended on Washington to provide instruments and relay data from western Canada. Circa 1939, more Canadian meteorologists belonged to the American Meteorological Society than to its much older British counterpart. Commercial aviation, in contrast, got its start with a multitude of tiny, independent airlines, often staffed by Canadian World War I veterans, who transported mail, payroll, and fire spotters beyond the reach of railways. These services played an important role in promoting export-led growth during the 1920s, stoking the fear that U.S. companies might try to move in to take advantage. In 1930, the increasingly independent Canadian Meteorological Service organized its first pilot balloon network to serve the growing list of Canadian airlines. This came to an abrupt halt in 1932 when the Great Depression compelled the Canadian government to cancel air mail subsidies as a cost-cutting measure. This nearly grounded commercial aviation in Canada.[51]

Lack of demand for aerological forecasts enabled a handful of Canadians, led by tropical meteorologist Andrew Thomson, to travel to Bergen at their own expense and learn "the latest methods in meteorology." Upon returning home, they began attempting air mass analysis. This required a great deal of trial and error as they developed new mapping techniques and an intuitive "feel" for the behavior of Canadian air masses, thus displaying the sort of local innovation that typically accompanies successful diffusion of scientific techniques. In 1933, Thomson established a graduate program in meteorology at the University of Toronto. Jacob Bjerknes and Sverre Petterssen both journeyed to Toronto for extended periods in the mid-1930s to help out their new converts. Meanwhile, Canadian and Irish politicians made an alliance with British Imperial Airways to forestall any attempt by Pan American Airways to dominate trans-Atlantic air service. This led to the creation of Trans-Canada Air Lines (TCA)—and to the importation of a fleet of Lockheed Electras and a team of former United Airlines executives from the United States to run the service. But thanks to Thomson's efforts, Canada did not have to import meteorologists to fill the new positions this demanded. Canadian scientists triumphantly produced forecasts using Bergen school techniques for the first experimental flights across the far northern Atlantic in 1935. By 1939, 34 of the 51 practicing meteorologists in Canada and Newfoundland had been trained by Thomson's program, nearly all of whom were employed by the Air Services Branch of the Dominion of Canada.[52]

Jacob Bjerknes spent the month of August 1939 lecturing at the Canadian Meteorological Office in Toronto. This series culminated with an unprecedented joint meeting of the Royal and American Meteorological Societies. Visiting dignitaries witnessed the official inauguration of the Canadian Branch of the RMS, the first autonomous branch to be organized in a British overseas dependency. The presence of a formidable line-up of Scandinavian scientists and their acolytes let everyone know how this long-anticipated event had come to pass (Figure 3). In fact, those in attendance could not avoid expressing a sense of ownership over these men: "I cannot regard Dr. Bjerknes as a Norwegian," declared one professor from the Imperial College of Science and Technology in London, "he is to me an international institution."[53]

These dignified ceremonies were followed up by serious atmospheric science, most notably during an all-day seminar on extra-tropical cyclone development led by Bergen school scientists. This seminar not only revealed how closely knit the Bergen school continued to be, it also showed how far its techniques had advanced. Those in attendance were honored with a detailed discussion of arguably the most significant advance in theoretical meteorology of the century: Rossby's derivation of his famous equation describing the propagation of large-scale atmospheric waves.[54] This meeting could not have confirmed more clearly the Bergen school's conquest of Canadian meteorology. Only one afternoon session granted significant time to scientists who were not declared adherents of the Bergen program. Canada's embrace of the "Norwegian School" during the 1930s thus helped it achieve scientific autonomy from both Great Britain and the United States.[55]

Unfortunately, the warm internationalist sentiments expressed at this reunion quickly faded. Germany invaded Poland two days after the meeting ended.

It is inaccurate to say that Scandinavian scientists were "stranded" in the United States by the outbreak of war, as is sometimes claimed. Several returned to Europe soon after, perhaps to their later regret. The United States had numerous attractions for those who remained. Most importantly, the core of this group had become highly dependent on each other for advances in their research. Rossby was able to derive his celebrated wave equation thanks to Jacob Bjerknes's progress in understanding upper-level conditions during "front-building." Bjerknes, in turn, had come to rely heavily on radiosonde observations, a new aerological technique using radio communication to transmit atmospheric conditions from a weather balloon to the ground. In Europe, this required delicate international coordination by the IMO Aerological Commission office in Berlin in the face of deteriorating political circumstances. The U.S. Weather Bureau, meanwhile, was well on its way toward converting its vast network to radiosonde; aerological observations on the scale of Europe's vast "experiment" of December 1937 were not far from becoming a daily occurrence in the United States.[56] North America also presented abundant opportunities for scientific leadership. In 1936, the

FIGURE 3   Canadian Meteorology Comes of Age: Participants at the first meeting of the Canadian Branch of the Royal Meteorological Society, 28–29 August 1939. "Norwegian School" scientists: 1. V. W. Ekman, 2. Jacob Bjerknes, 3. Halvor Solberg, 4. Sverre Petterssen. Not pictured: C-G Rossby. Prominent North American disciples: a. Horace R. Byers, b. Bernhard Haurwitz (a recent immigrant from Germany), c. Francis W. Reichelderfer, d. Andrew Thomson. Many of those pictured received training in Bergen School techniques at the University of Toronto. Source: *Quarterly Journal of the Royal Meteorological Society* 66, supplement (1940).

patriarch of Norwegian oceanography, Bjørn Helland-Hansen, had arranged for Harald Sverdrup to become director of the Scripps Institution of Oceanography (SIO). This established a firm beachhead for the Bergen school on the Pacific coast. Like Rossby before him, Sverdrup seized this opportunity to leave Norway for a position where he could "plan and have an influence on others' work" rather than live under the shadow of the founders of the Bergen school—in spite of the intense homesickness this caused him and his wife.[57]

With Europe so unsettled, a competition ensued to see where Jacob Bjerknes would land, establishing one of the world's centers for meteorological research in the process. All of this was well underway before a fleet of Junkers JU-52 hydroplanes led the German invasion of Norway on 9 April 1940 and ended its reign as the headquarters of the Bergen school.[58] Attracted by Sverdrup's presence, Bjerknes soon settled on the University of California's southern branch in the city of Los Angeles, one of the world's emerging centers for scientific and technological research and its largest producer of aircraft.[59] Bjerknes enthusiastically embraced the scientific challenges posed by Pacific Coast meteorology, helped along by timely U.S. Weather Bureau grants to study "typical frontal and air mass patterns over the Pacific, with particular reference to . . . the general circulation . . . and flood-producing rains in California."[60] But he held out signing a contract until he negotiated the hire of meteorologist Jörgen Holmboe, a long-time Bergen colleague. Holmboe, in turn, extracted a promise from Bjerknes to settle permanently in Los Angeles. Like so many settler colonists, these Scandinavians ultimately preferred to live where they already had friends from the home country. Pull factors far outweighed push factors in bringing them to the United States.[61]

But these men had little time for research, as they became consumed with planning and providing hands-on "Education and Training in Meteorology for the Armed Forces." Sverdrup, in fact, spent far more time than he intended teaching oceanography to UCLA meteorology students—to the lasting benefit of the meteorology profession—when "remote hearsay" regarding supposed German sympathies delayed his security clearance for wartime research.[62] As Roger Turner explains elsewhere in this volume, these efforts completed the colonization of North America by Scandinavian geophysical science: During the course of the war, 1,400 students at UCLA and SIO, 375 students in Canada, hundreds of practicing Weather Bureau scientists, and thousands of students attached to the British Commonwealth and U.S. armed forces received "advanced" training at programs run by disciples of the Bergen school.[63] These educational programs provided a launching point for the spread of polar front theory to the rest of the Western Hemisphere.

NORDIC SCIENCE INVADES THE TROPICS

The westward shift of the "Norwegian School" changed its geographical perspective. The leaders of the Bergen school could not forget events in Europe, of course. Through a long-time German friend, Jacob Bjerknes worked to assure good treatment for his father and to prevent the execution of a Bergen meteorologist accused of espionage in German-occupied Norway.[64] Sverdrup provided jobs for refugee scientists such as Wladyslaw Gorczynski, a prominent Polish climatologist with years of experience directing solar radiation research in the tropics, even though Sverdrup, reflecting the Bergen school's biases, had *"no intention whatsoever* to add climatology to the subjects under investigation at the Scripps Institution."[65] With European scientific channels closed off, these Scandinavian expatriates began turning their gaze south toward Latin America.

A major international scientific event helped inspire this shift. In May 1940, a vast contingent of Latin American scientists and diplomats descended on Washington, D.C., for the Eighth American Scientific Congress, the most recent in a series dating back to Buenos Aires in 1898 and Montevideo in 1901. President Franklin Roosevelt set the tone with an important statement of U.S. policy regarding hemispheric unity: "The annihilation of time and space" by "planes and bombs" meant that "every acre—every hectare—of the Americas" was in danger. "I am a pacifist," Roosevelt declared, making reference to his Good Neighbor Policy rejecting U.S. military intervention and embracing bilateral collaboration in the Americas. "You, my fellow citizens of twenty-one American Republics are pacifists. But I believe . . . you and I . . . will act together to protect and defend by every means our science, our culture, our freedom and our civilization." A parade of Latin American spokesmen echoed Roosevelt's "Messianic words" affirming the "duty of science" to make "Pan Americanism . . . a living and tangible thing."[66]

About 125 people attended the meteorological session of the Congress, though only one Latin American meteorologist read his own paper of the 36 submitted to the session. The Chief of the Mexican Meteorological Service impressed everyone with his report detailing Mexico's upper-air weather observatories and radio network, all modeled on U.S. Weather Bureau operations.[67] The American Meteorological Society made sure this meeting had a broad impact by distributing its published proceedings as a supplement to the AMS *Bulletin*. At the urging of Argentina's chief meteorologist Alfredo Galmarini, the foremost proponent of scientific internationalism among South American atmospheric scientists, the AMS Council took a further step: In 1941, it decided the "Society should fulfill its *true hemispherical nature* by publishing parts of its BULLETIN in Spanish as well as English" in order to encourage "greater interchange of ideas and closer personal relations between Spanish, Portuguese and English speaking Americans." This ran counter to well-established plans to publish "an Inter-American review of meteorology" edited by Galmarini's Uruguayan rival, Luis Morandi—and it was likely intended by Galmarini to counter-balance German influence in Argentine science.[68]

This incident hints at far deeper conflicts dividing the region, many focused on air power. Bolivia had just lost its protracted "Chaco War" with Paraguay, notwithstanding the contribution of LAB's growing fleet of Junkers aircraft. In 1938, Brazil fired off a salvo of threats after learning that the U.S. Army had made a secret arrangement to re-staff Argentina's flight-training program after Germany pulled out. And in 1941, a coordinated blitzkrieg offensive by Peru seized a vast portion of Ecuador's Amazonian frontier. Not only were South America's militaries mutual enemies, they were internally fractured into pro-French, German, and U.S. cliques. The situation became so tense that Panair do Brasil's chief executive was hauled before a military tribunal for German espionage in 1942.[69]

Aviation technology and atmospheric science both played critical roles in achieving a semblance of hemispheric political unity during World War II. As we have seen, German science and industry exercised substantial influence over aviation and meteorology in South America, mainly because of relationships established during the 1920s long before the National Socialists seized power. But U.S. observers, as a reflection of their own neocolonial aspirations, self-servingly interpreted these relationships as evidence of a "Nazi imperialist drive to dominate South American airlines."[70]

An ill-timed request by Deutsche Lufthansa's Ecuadorian subsidiary to initiate service to the Galápagos, together with Nelson Rockefeller's influential report calling for the "economic defense" of the Americas, triggered an aggressive U.S. campaign to eliminate even the weakest German ties to Latin American aviation. On 8 June 1940, under intense pressure, Pan American Airways summarily fired 80 pilots, mechanics, and administrators from SCADTA—some with over one million hours of flight experience.

Many German- and Austrian-born employees had formed local families and taken Colombian citizenship, including Peter Paul von Bauer, who reputedly did so as a reaction against Nazi overtures. Some ended up among the 4,058 Germans, 2,264 Japanese, 258 Italians—and not a few Jews and communists—who were deported from Latin America to U.S. prison camps for the duration of the war. In 1941, the Peruvian, Bolivian, and Ecuadorian governments expropriated local German airlines, to the direct benefit of Pan American-Grace Airways (Panagra) which immediately took over their routes. Los Angeles manufacturers also benefited from "de-Germanification." With help from an $8 million U.S. program, several airlines acquired fleets of new Lockheed planes to replace old Junkers aircraft. But nationalists in the Brazilian military, fearful of the growing strength of Pan American Airways, stubbornly resisted U.S. efforts to close down Syndicato Condor. After years of collaborating with Germany on development issues, Getúlio Vargas eventually gave way to U.S. pressure, though only after finalizing arrangements for building a string of air bases through northern Brazil, a vast coal-steel complex, and other lucrative concessions. In return, the U.S. military gained access to northern Brazil for the duration of the war. These agreements entailed the installation of U.S. Army Air Force forecasters—most with training in polar front theory—along the airway connecting Florida to North Africa via Brazil. This included placing a U.S. supervisor in charge of Rio de Janeiro's aviation forecast center, the Deutsche Seewartes's former stronghold. Syndicato Condor became Brazilian property and its remaining German employees were fired when Brazil declared war on Germany in August 1942. Its new management was promptly rewarded with four coveted DC-3s, the first of an eventual 278 DC-3/C-47s purchased by major Brazilian airlines. Vargas hailed these agreements as symbols of Brazil's economic emancipation.[71]

In one notorious case, the United States used naked force to move this policy forward. On 16 January 1942, under the cover of night and a secret diplomatic agreement, U.S. military detachments invaded three of the Galápagos Islands and the tip of the oil-rich Santa Elena peninsula in Ecuador in order to construct bases for the defense of the western approach to the Panama Canal. With much of Ecuador under Peruvian occupation, its government was in no position to protest. In return for a mere $35,000, it signed an agreement after the fact granting a long list of demeaning concessions.[72] The U.S. Sixth Weather Squadron used this opportunity to install the first three meteorological observatories with radiosonde capability to be established on the Pacific coast of South America. Its officers acquired aerological data of unprecedented detail for the "tropical front" thought to separate air masses of the northern and southern hemisphere in the course of thousands of survey flights between Guatemala, the Galápagos, and the South American mainland (Figure 4).[73] But the U.S. military ignored repeated entreaties to share these observations with allied Ecuadorian and Mexican meteorologists. Nor did they "put themselves into contact with the Meteorological Service of Ecuador so as to find a way to maintain the functioning of this important center . . . in the equatorial zone of the Pacific Ocean" at the end the war—to the eternal regret of scientists interested in the El Niño phenomenon.[74] This U.S. military foray in Ecuador did nothing to unite Latin American enemies in a common cause.

In stark contrast, civilian atmospheric scientists in the United States embraced Nelson Rockefeller's emphasis on peaceful, international cultural and scientific exchange with Latin America.[75] Jacob Bjerknes headed straight to Gorczynski's old stomping ground, Mexico City, between his first two semesters at UCLA in order to teach air mass analysis to the Mexican Meteorological Service. The U.S. Weather Bureau then sent its chief scientist R. Hanson Weightman—Rossby's first American disciple—on a tour of South America "to attain . . . greater collaboration in the exchange of meteorological information" and to extend invitations to a hemispheric meteorological meeting to be held in Washington. The Weather Bureau also put together a six-month training program for ten circum-Caribbean students focused on hurricane prediction and radiosonde technique.[76]

In the fall of 1941, Rossby hatched a plan to expand these activities. He suggested an extended tour of the "principal meteorological centers" of Latin America by two "recognized meteorologists" with "tact and sympathy for the Latin American temperament." Rossby also proposed an exchange program that would have sent ten young U.S. meteorologists to Latin America to equip five new radiosonde observatories, and ten young South American nationals to the United States for advanced meteorological training. He explicitly hoped to extend "modern methods" of air mass analysis to temperate regions of the Southern Cone and to find out if Bergen school techniques could provide insights into the vertical stability and moisture content of "true equatorial air." The fact that Rossby volunteered himself and Bjerknes to serve as missionaries shows the high priority he placed on this plan's success—it meant the conquest of a vast new region for the Bergen school. Others immediately began promoting Rossby's plan as a means to eliminate Germans from South America's weather services and prevent "Nazi sabotage" of hemispheric forecasts.[77]

The United States' full mobilization after Pearl Harbor prevented Rossby and Bjerknes from taking this tour. Instead, they got something far better: an elaborate Latin American training program that lasted for the duration of the war. The U.S. Weather Bureau kicked off this program with the Inter-American Institute of Meteorology held in Medellín, Colombia, in 1943. A total of 200 students representing every Latin American republic attended this six-month course. Nelson Rockefeller's Interdepartmental Committee for Cultural and Scientific Cooperation hastily translated a U.S. textbook intended for commercial pilots for the Institute. It focused explicitly on "polar front theory" and "air mass analysis" as originated by "Norwegian scientists," but unlike many textbooks of that time, its examples derived almost entirely from the United States. The Institute depended on Latin American assistant instructors to provide regional applications.[78] Students also received practical training in map analysis, tracking pilot balloons, and the English language. Forty-six graduates with the best language and mathematical skills were then sent to the United States for nine to twelve months of advanced training.[79] California's long-standing efforts to promote itself as Latin America's new Mediterranean metropole paid off handsomely in this case.[80] In view of the "long background of connections . . . this part of the country has with Latin-American affairs," 28 members of this initial advanced class were sent to UCLA, 42 more by the end of the war.[81]

Latin Americans enthusiastically embraced the opportunities presented by this training program. Eight hundred applicants competed for the original 200 positions at the Institute. Many were quickly rewarded for their efforts. Approximately half of the Institute's graduates promptly obtained professional positions in meteorology. Gustavo Wray, for example, became Panagra's chief observer at the Guayaquil airport, and he was eventually charged with supervising Ecuador's entire coastal network of government observatories. Rafael Dávila Cuevas, the enterprising son of highland immigrants to Lima, represented a new class "of strong, dark complexion" that had only recently been allowed to attend Latin America's universities. When the war forced him to abandon geophysical studies at the University of Kyoto, he headed straight for UCLA to work with Bjerknes. As a professor of geophysics at the Universidad Nacional Mayor de San Marcos, Dávila went on to train a new generation of Peruvian scientists in atmospheric science—on a couple occasions, with Bjerknes's direct assistance.[82] In addition to serving individual ambitions, these educational exchange programs enabled Latin Americans, like Canadians, to fulfill growing regional demand for meteorological expertise, and they helped several nations' efforts to form a technical elite. In the process, they established direct ties of allegiance between the Latin American and U.S. meteorological professions. Programs like these worked so well toward these ends that they soon formed the backbone of Cold War development strategies in the Third World.[83]

These high-profile international programs provided direct stimulus for the reorientation of atmospheric science in much of Latin America. Brazil, Mexico, and Ecuador dutifully initiated detailed investigations of regional air masses and established training programs of their own to produce experts in frontal analysis.[84] To better serve regional aviation, Ecuador and Bolivia abruptly shifted their meteorological services away from climatology and toward synoptic meteorology during the war. The Ecuadorian government also initiated the expensive task of retrofitting its observatories with U.S. equipment—though not fast enough to make up for the systematic breakdown of German and French precision instruments lacking recording drums and routine factory maintenance during the war.[85] Luis Mena, an instructor at Medellín, returned home intent on initiating pilot balloon soundings at the Quito Astronomical Observatory. Obtaining local government support was the least of his problems. U.S. military staff controlled the wartime supply of meteorological instruments, and they placed Ecuador near the bottom of the waiting list. (After all, the U.S. Army Air Force operated its own observatories in Ecuador.) Only direct intervention by the head of the U.S. Weather Bureau to provide a package of scarce balloon valves enabled the Quito Observatory to initiate upper-air observations in March 1946 after a two-year wait.[86] This sort of treatment became a major sore spot in hemispheric relations. After making big promises to Latin America in order to cement the wartime alliance, the United States repeatedly required its southern allies to wait in line for basic economic and technical aid after the war: first until after the rebuilding of Europe and Japan under the Marshall Plan, then until after the Korean conflict, then until after international organizations took the lead. (An all-out revolt in a U.S. client state, the 1959 Cuban Revolution, eventually forced a change in U.S. development policy and scientific engagement in the region.)[87]

Meanwhile, the reorientation of scientific resources toward synoptic meteorology and other wartime disruptions discouraged some promising lines of climatological research in the Southern Hemisphere. Before the war, the main observatory in Santiago de Chile had nurtured scientific ties that spanned the entire Pacific. It took part in a project with New Zealand and Argentina to determine the average trajectories of depressions and anticyclones in the South Pacific. Like their colleagues in India, Chilean meteorologists showed a particular interest in the correlation between abnormal precipitation and the Southern Oscillation; to this end, they carefully collected daily reports transmitted by shortwave from Manila, Hanoi, and Batavia.[88] Ecuadorian scientists were similarly engaged in projects to relate solar radiation, abnormal precipitation patterns, and variations in the strength of the "El Niño countercurrent" off the coast of Ecuador. They had even established a transoceanic data exchange that used high rainfall anomalies in coastal Ecuador and Peru to produce drought forecasts for the Dutch East Indies. World War II not only cut off Quito from Batavia, anti-espionage regulations also prevented Ecuadorian government meteorologists from freely exchanging data with private companies operating locally—a vital source of data from the El Niño-prone coast.[89] As Harald Sverdrup's promise to Wladyslaw Gorczynski illustrates, Bergen school scientists viewed research focused on statistical weather cycles and solar forcing with open disdain. Their overwhelming triumph during World War II converted these approaches into a signifier of scientific backwardness. These interests did not revive until after the International Geophysical Year of 1957–1958.[90]

THE BIRTH OF MODERN TROPICAL METEOROLOGY

In the midst of all this, some disciples of the Bergen school became convinced that polar front theory had almost nothing to contribute to tropical forecasting. In 1943, through Rossby's initiative, the University of Chicago organized an Institute of Tropical Meteorology at the University of Puerto Rico. Seven of the original 46 Latin American students who came to the United States for advanced training were invited to this Institute. Rossby realized that his growing meteorological empire would never last—and wartime forecasts in the tropics would fail—unless they were founded on rigorous empirical research based in the Torrid Zone. He selected Puerto Rico, not only because it was close by and governed directly by the United States, but also because U.S. and Puerto Rican scientists had compiled the oldest, most detailed set of pilot balloon observations for this hurricane-prone region. He looked far afield for an experienced tropical meteorologist to direct the Institute: Clarence Palmer, a New Zealander who had become a disciple of the Bergen school during Jörgen Holmboe's extended stay at the New Zealand Meteorological Office in the mid 1930s. Palmer had gained renown as a tropical forecaster during his tour of duty at Guadalcanal with the Royal New Zealand Air Force. Rossby also arranged for Herbert Riehl, a promising young German immigrant prohibited from more sensitive tasks, to attend as a student.[91]

Palmer and Riehl dedicated themselves to the systematic study of upper-air data from the Caribbean, including high-quality radiosondes that had recently become available from U.S. colonial observatories at San Juan, Guantánamo Bay, and Swan Island. They relied on a number of techniques developed by Scandinavian meteorologists, including Vilhelm Bjerknes's method for analyzing streamlines and Rossby's kinematic approach to atmospheric waves. But empirical evidence forced them to reject the principle that frontal dynamics at the boundary between air masses contributed significantly to tropical weather. Riehl, instead, concentrated on the development of perturbations *within* tropical air masses, particularly "waves in the easterlies." These westward-moving troughs of low pressure not only provided a valuable tool for predicting convective rainfall and upper-level winds in the Caribbean, they also marked where a hurricane was likely to form (Figure 4).[92]

In a review of "progress" made by the profession during the 1940s, Palmer could not have been more pessimistic about the future of the air-mass approach to tropical meteorology. "This theory failed to pass the test to which all theoretical work in meteorology must finally come; . . . it was found almost useless as a guide to short-period forecasting in low latitudes." In fact, "a large part of the more successful short-range tropical forecasting in World War II," he concluded, had relied on older, statistical approaches to climate—exactly the kind of meteorology the triumph of the Bergen school had done so much to discourage. Pilot balloons, as it turned out, proved ill-suited to producing reliable data on upper-level winds in the humid tropics except during fair weather—they usually got lost in the convective cloud cover associated with tropical precipitation. Only expensive, high-precision barometers and barographs, radiosondes, airplane surveys, and radar produced meaningful measures of day-to-day changes governing the tropical atmosphere. These realizations led to "the disillusionment of a whole generation of tropical meteorologists."[93] This was particularly true in countries like Ecuador and Peru where the resulting "lack of knowledge of the meteorological processes involved in the formation of daily weather" and lack of resources to replace low-precision instruments deprived southern forecasters of the basic theoretical and empirical tools they needed to do

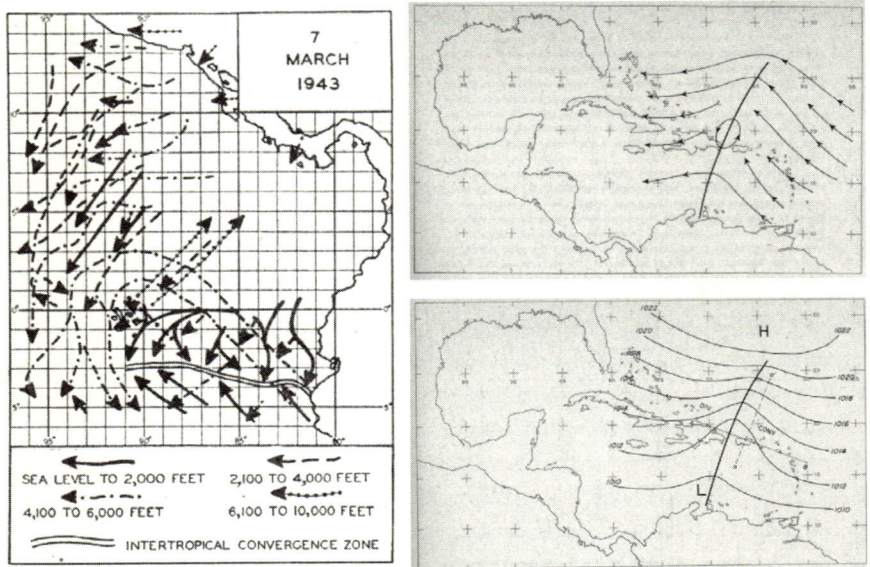

FIGURE 4   Two competing models of tropical meteorology at the end of World War II.

Left: The Equatorial Front. Following the lead of Vilhelm Bjerknes, tropical fore-casters used the wind field in the tropics to map the boundary between tropical air masses, even though marked temperature differentials rarely accompanied these "fronts." This map of winds over the eastern tropical Pacific shows the "inter-tropical convergence zone" several degrees south of its usual position in March, a hallmark of the El Niño phenomenon, as well as the presence of "anti-trades" above 6,100 feet over the Galápagos.

Right: The Easterly Wave. In these idealized diagrams, upper-level winds (above, at 10,000 feet) and surface-level pressure (below) indicate a downward sloping atmospheric wave transecting the Antilles from north to south. By locating "waves in the easterlies," a synoptic meteorologist could predict zones of convective storms (marked by dotted line) and "shear zones" (abrupt changes in wind direction). Most importantly, a forecaster could expect these waves to propagate east to west and sometimes give birth to tropical depressions.

Sources: Leo Alpert, "Weather over the Tropical Eastern Pacific Ocean, 7 and 8 March, 1943," *Bulletin of the American Meteorological Society* 27 (1946): 389; Herbert Riehl, *Waves in the Easterlies and the Polar Front in the Tropics*, University of Chicago, Department of Meteorology, Miscellaneous Reports No. 17 (Chicago: University of Chicago Press, 1945), 7, 9.

reliable work.[94] (This discouraging situation mirrored the postwar circumstances that faced Latin American economists trained in the North. They responded by generating their own theoretical approach to economic reality: dependency theory.)[95] South of the Rio Grande, only the meteorology profession in Argentina—the only Latin American state that stayed out of the wartime alliance—seems to have benefited unambiguously from the war. The knowledge young Argentine meteorologists gained at Medellín and UCLA worked quite well in the temperate zone of South America. After the war, a number of gifted German scientists immigrated to Argentina. Like Scandinavian scientists, they, too, welcomed the opportunity to pursue upper-air research while living in settler colonies of their compatriots.[96]

Tropical meteorology in the United States had a very different fate. As a result of Palmer and Riehl's research, Rossby abruptly revised his understanding of the role of the tropics in the general circulation after the war, and he used his enormous influence to encourage the development of a new Tropical Meteorology. After four years in Puerto Rico, Herbert Riehl returned to the University of Chicago where he completed a Ph.D. under Rossby's supervision. With Rossby's support, Riehl stayed on in Chicago and established a major research school dedicated to the study of tropical cyclones. Clarence Palmer joined the new Institute of Geophysics at UCLA built around Jacob Bjerknes after the war. In this way, Chicago, Los Angeles, and later Miami emerged as major centers of action for tropical research and utterly eclipsed centers like Havana, Manila, and Batavia that had flourished under colonial rule.[97]

What, then, does this case study tell us about the historical relationship between science, technology, and neocolonialism? The protagonists of this hemispheric struggle over airways were conscious that these forces were closely intertwined and mutually reinforcing. But different actors used a wide variety of strategies in an attempt to manipulate these relationships to their benefit. Some of the strategies formed patterns that might be considered "national styles in science."[98] The contrast between German and U.S. "styles" was particularly marked: Germans often used science to lead the way for other neocolonial endeavors, while American science tended to follow on the heels of other forms of dominance, often imposed from the top-down. Scandinavians used the struggle over airways to assert their supremacy more narrowly, in a scientific realm; their endeavors, meanwhile, helped the Bergen school to maintain a lasting sense of cohesion, even when the polar front paradigm revealed its many limitations. Canadian scientists, in turn, used

Scandinavian supremacy to establish professional autonomy from the U.S. and U.K. This diversity of strategies was just as marked within Latin America. Argentines, Brazilians, and Colombians repeatedly worked conflicting neocolonial aspirations in South America to their own advantage. Some strategies did not work: Ecuadorians, despite their best efforts, found themselves in a position inferior to Puerto Rico and the Canal Zone, at least where scientific and technological development were concerned, though they did convince the U.S. to abandon its military bases after the war.

The diffusion of atmospheric science and aviation technology between 1919 and 1945, in the final analysis, played an important role in dividing the Americas into First and Third Worlds and exacerbated regional disparities of wealth, productivity, expertise, and innovative capacity. Most of these outcomes were unintended, and some were almost incomprehensible to those involved. How else can we explain the widespread, North-South consensus that emerged after the war promoting further education, scientific research, and technological innovation as answers to these "developmental problems"? Perhaps neocolonialism deserves its reputation as "the worst form of imperialism," not because of the economic constraints it created, but because of the "particular cast of mind" it encouraged, in which scientific knowledge, technological development, and economic growth came to be seen as automatically beneficial by so many postcolonial rulers, despite their long-term consequences for local environments and livelihoods.[99]

ACKNOWLEDGMENTS

I am greatly indebted to the editors of this volume and numerous friends and colleagues who read early versions of this work, especially Byron Caminero-Santangelo, Jorge Cañizares, Katherine Clark, Megan Greene, Bruce Hunt, Henning Krause, Liz MacGonagle, Andrew Paxman, Leslie Tuttle, Kim Warren, Nathan Wood, the History of Science Colloquium at the University of Oklahoma, my own research seminar on the Colonial and Postcolonial World, and most of all, to Mirna Cabrera. The American Meteorological Society, University of Kansas Center for Research, and KU Office of International Programs provided research support, particularly for travel. Ericson López Izurieta, director of the Escuela Politécnica National's program in astrophysics, and his Staff provided indespensible access to the archives of the Observatorio Astronónomico de Quito.

NOTES

Abbreviations used in notes: Archivo del Observatorio Astronómico de Quito: Comunicaciones Recibidas (AOAQ-CR); U.S. Weather Bureau correspondence (AOAQ-USWB); International Meteorological Organization correspondence (AOAQ-IMO). University of California, Los Angeles, Charles E. Young Research Library: Jacob Bjerknes papers, special collection 1709 (UCLA-Bjerknes). UCLA Archives, Office of Chancellor, Administrative Files, record series 359: Scripps Institution of Oceanography, General Matters (UCLA-SIO); Department of Physics, Meteorology Program, (UCLA-PhysMet); Department of Meteorology (UCLA-Met). Scripps Institution of Oceanography Archives, University of California, San Diego: Office of the Director, Sverdrup, Records, (SIO-Sverdrup). *Quarterly Journal of the Royal Meteorological Society (QJRMS)*; *Bulletin of the American Meteorological Society (BAMS)*.

1. D. A. Brading, *The First America: The Spanish Monarchy, Creole Patriots, and the Liberal State, 1492–1867* (Cambridge: Cambridge University Press, 1991), 614–615; Frank Safford, *The Ideal of the Practical: Colombia's Struggle to Form a Technical Elite* (Austin: University of Texas Press, 1976), esp. 101–102. See also Mauricio Tenorio-Trillo, *Mexico at the World's Fairs: Crafting a Modern Nation* (Berkeley and Los Angeles: University of California Press, 1996); Miguel Angel Centeno, "The New Leviathan: The Dynamics and Limits of Technocracy," *Theory and Society* 22, no. 3 (June 1993): 307–335; John Markoff and Verónica Montecinos, "The Ubiquitous Rise of Economists," *Journal of Public Policy* 13, no. 1 (1993): 37–68.
2. Gregory T. Cushman, "Viñes, Benito," in *New Dictionary of Scientific Biography*, ed. Noretta Koertge (New York: Charles Scribner's Sons, in press); Erik Larson, *Isaac's Storm: A Man, a Time, and the Deadliest Hurricane in History* (New York: Vintage, 2000).
3. For recent reviews, see "Focus: Colonial Science," *Isis* 96, no. 1 (2005): 52–87; Roy MacLeod, ed., "Nature and Empire: Science and the Colonial Enterprise," *Osiris* 2nd ser., vol. 15 (2000). On technology, see founding works by Daniel R. Headrick, *The Tools of Empire: Technology and European Imperialism in the Nineteenth Century* (New York: Oxford University Press, 1981); Headrick, *The Tentacles of Progress: Technology Transfer in the Age of Imperialism, 1850–1940* (New York: Oxford University Press, 1988).
4. The two best-known interpretive models for understanding the global spread of Western science were heavily based on the experience of British settler colonies: George Basalla, "The Spread of Western Science," *Science* 5 May 1967: 611–622; Roy MacLeod, "On Visiting the 'Moving Metropolis': Reflections on the Architecture of Imperial Science," *Historical Records of Australian Science* 5, no. 3 (Nov. 1982): 1–16. For critiques of these models, motivated by the histories of Canada and Latin America, see Richard A. Jarrell, "Differential Development

and Science in the Nineteenth Century: The Problems of Quebec and Ireland," in *Scientific Colonialism: A Cross-Cultural Comparison* (Washington, DC: Smithsonian Institution Press, 1987), 323–350; Antonio Lafuente and María L. Ortega, "Modelos de mundialización de la ciencia," *Arbor* 142, no. 558–560 (June-Aug. 1992): 93–117.

5. Shiv Visvanathan, "On the Annals of the Laboratory State," in *Science, Violence & Hegemony: A Requiem for Modernity*, ed. Ashis Nandy (New Delhi: Oxford University Press, 1990), 257–288; cf. David Wade Chambers and Richard Gillespie, "Locality in the History of Science: Colonial Science, Technoscience, and Indigenous Knowledge," *Osiris*, 2nd ser., vol. 15 (2000): 221–240.

6. This discussion was strongly influenced by Robert J. C. Young, *Postcolonialism: An Historical Introduction* (Malden, MA: Blackwell Publishers, 2001), ch. 4; also by Jürgen Osterhammel, *Colonialism: A Theoretical Overview*, trans. Shelley L. Frisch (Princeton, NJ: Markus Wiener Publishers, 1997).

7. Tulio Halperín Donghi, *The Contemporary History of Latin America*, trans. John C. Chasteen (1969; Durham, NC: Duke University Press, 1993), ch. 4–5.

8. John Herd Thompson and Allen Seager, *Canada 1922–1939: Decades of Discord* (Toronto: McClelland and Stewart, 1985), esp. 190–192.

9. Fernando Henrique Cardoso and Enzo Faletto recognized, long ago, the agency of locals in forming relations of dependency, even as they emphasized the structural factors limiting their choices, in *Dependency and Development in Latin America*, trans. Marjory Mattingly Urquidi (Berkeley and Los Angeles: University of California Press, 1979).

10. See Peter Fritzsche, "Machine Dreams: Airmindedness and the Reinvention of Germany," *American Historical Review* 98, no. 3 (June 1993): 685–709; Paul Forman, "Scientific Internationalism and the Weimar Physicists: The Ideology and its Manipulation in Germany after World War I," *Isis* 64, no. 2 (June 1973): 151–180.

11. This study is based on multilingual research on three continents, including scientific archives in Southern California, library materials from the Ibero-Amerikanisches Institut in Berlin, and a remarkable archive of meteorological correspondence from Ecuador.

12. Marietta Benkö and Bernhard Schmidt-Tedd, *Historical Manuscripts [in] Relation to the Invention of the First Balloon and Its Maiden Flight in Lisbon, August 8, 1709* (Cologne: Institute of Air and Space Law at Cologne University, 2002); Paul Hoffman, *Wings of Madness: Alberto Santos-Dumont and the Invention of Flight* (New York: Theia, 2003), ch. 13; Gary Glen Kuhn, "The History of Aeronautics in Latin America" (Ph.D. diss., University of Minnesota, 1965), ch. 1, 5.

13. The main criticism of diffusionist thinking stems from its association with simplistic, unidirectional theories used to legitimate colonial rule; see J. M. Blaut, *The Colonizer's Model of the World: Geographical Diffusionism and Eurocentric History* (New York: Guilford Press, 1993).

14. Kuhn, "The History of Aeronautics in Latin America," 36–42, 46–48, 51–54, 65–66, 74–79, 120–124, 160–161, 165; Vicente Gesualdo, "Andar en globo: Las primeras ascensiones aerostáticas en Argentina y Sudamérica," *Todo es historia* (Buenos Aires) Jan. 1983: 40–47, 54–61.

15. The oft-repeated phrase "Transportation Is Civilization" was the motto of the technocratic Aerial Board of Control in Rudyard Kipling's futuristic short story "With the Night Mail: A Story of 2000 AD," (1904); Halperín, *Contemporary History of Latin America*, ch. 4–5.

16. For a detailed chronology of Hargrave's contributions, see John Bird and Russell Naughton, "Lawrence Hargrave: Australian Aviation Pioneer," Monash University, http://www.ctie.monash.edu.au/hargrave/ (accessed 30 May 2005); Kh. Khrgian, *Meteorology: A Historical Survey,* 2nd ed., trans. Ron Hardin (1959; Jerusalem: Israel Program for Scientific Translations, 1970), 275–297.

17. *Segundo reunión del Congreso Científico Latino Americano celebrada en Montevideo del 20 a 31 de marzo de 1901* (Montevideo: Tip. y Enc. "Al Libro Inglés," 1901), xiii, 13, 19–20; L. Moranci, "Los primeros sondeos aerológicos en el Uruguay," *Revista meteorológica* (Montevideo) 1, no. 1 (Jan. 1942): 34–39; J. Patterson, "A Century of Canadian Meteorology," *QJRMS* 66, supplement (1940): 26–28.

18. David E. Omissi, *Air Power and Colonial Control: The Royal Air Force, 1919–1939* (Manchester: Manchester University Press, 1990); R. E. G. Davies, *A History of the World's Airlines* (London: Oxford University Press, 1964), esp. ch. 2, 4, 6, 13; Wesley Phillips Newton, *The Perilous Sky: U.S. Aviation Diplomacy and Latin America, 1919–1931* (Coral Gables, FL: University of Miami Press, 1978), 22–24; José Ignacio Forero F., *Historia de la aviación en Colombia* (Bogotá: Aedita Editores, 1964), 142–145. On the diffusion of European military influence in Latin America, see Frederick M. Nunn, *Yesterday's Soldiers: European Military Professionalism in South America, 1890–1940* (Lincoln: University of Nebraska Press, 1983); and William F. Sater and Holger H. Herwig, *The Grand Illusion: The Prussianization of the Chilean Army* (Lincoln: University of Nebraska Press, 1999).

19. J. de Sampaio Ferraz, *Meteorologia brasileira (Esboço elementar de sues principaies problemas)* (São Paulo: Companhia Editora Nacional, 1934), 26–27, 136; Guillermo Hoxmark, *La aviación y la meteorología* (Buenos Aires: Oficina Meteorológica Nacional de la República Argentina, 1924), 3, 13.

20. My interpretation relies heavily on Stefan H. Rinke, *"Der letzte freie Kontinent": Deutsche Lateinamerikapolitik im Zeichen transnationaler Beziehungen, 1918–1933*, 2 vols. (Stuttgart: Heinz, 1996).

21. Rinke *"Der letzte freie Kontinent"*, 2:659–668; R. E. G. Davies, *Airlines of Latin America since 1919* (Washington, DC: Smithsonian University Press, 1984), 207–216, 628–629; Herbert Boy, *Una historia con alas*, 2nd ed. (Bogotá: Editorial Iqueima, 1963), 80–85.

22. Rinke, "Die Firma Junkers auf dem lateinamerikanischen Markt, 1919–1926," in *Grenzenlose Märkte?: Die deutsch-lateinamerikanischen Wirtschaftsbeziehungen vom Zeitalter des Imperialismus bis zur Weltwirtschaftskrise*, ed. Boris Barth and Jochen Meissner (Münster: Lit Verlag, 1995), 157–184; Rinke, *"Der letzte freie Kontinent"*, 1:35–36, 2:684–699; Amalia Villa de la Tapia, *Alas de Bolivia: Síntesis histórica de la aviación nacional* (La Paz: Editorial Aeronáutica Fuerza Aérea Boliviana, 1974), 1:141–165.

23. Gustavo Arias De Greiff, *Otro cóndor sobre los Andes: Historia de la navegación aérea en Colombia* (Bogotá: BANCAFE/FIDUCAFE, 1999), 110–113; Boy, *Una historia con alas*, 86–88; Florencio Gómez Núñez, *Mis apuntes sobre la aviación venezolana: Documento para la historia de al aviación venezolana* (Caracas: Impresos Moranduzzo, 1970), 153–158.

24. Willy Rudloff, "Ozeanflugwetterdienst," in *Schiffahrt und Meer: 125 Jahre maritime Dienste in Deutschland*, ed. Peter Ehlers, Georg Duensing, and Günter Heise (Berlin: Verlag E. S. Mittler, 1993), 144, 153.

25. Mario de Montemar, "El nuevo servicio de aerología en el Observatorio Nacional," *Bohemia* 14 Feb. 1926: 14; José Carlos Millás, "Iniciación de los estudios aerológicos en Cuba," *Boletín del Observatorio Nacional* (Havana) 22, no. 1 (1926): 3–12 [quote p. 8]; Millás, "Observaciones aerológicas," *Boletín del Observatorio Nacional* 22, no. 9 (1926): 173; Cushman, "Viñes, Benito."

26. Newton, *The Perilous Sky*, 64–70; Davies, *Airlines of Latin America*, 338; Rinke, *"Der letzte freie Kontinent"*, 2:668–672; Rudloff, "Ozeanflugwetterdienst," 156; Walter Georgii and Heinrich Seilkopf, "Ergebnisse einer flugwissenschaftlichen Forschungsreise nach Columbia (S.A.)," *Aus dem Archiv der Deutschen Seewarte* (Hamburg) 43, no. 3 (1926): 5–10, 14, 52–56; Herbert Riehl, *Tropical Meteorology* (New York: McGraw-Hill, 1954), 243–244.

27. Newton, *The Perilous Sky*, 70–77; see also Nancy Mitchell, *The Danger of Dreams: German and American Imperialism in Latin America* (Chapel Hill: University of North Carolina Press, 1999).

28. Davies, *Airlines of Latin America*, 217–224, 338–344, 347–350; Rudloff, "Ozeanflugwetterdienst," 156.

29. This was the second time an accident prevented Argentina from establishing a meteorological presence on the southern continent. Antonio Pauly, *Proyecto de Expedición Aérea Argentina al Polo Sur* (Buenos Aires, 1926), esp. 37–39, 49–51; Alfredo G. Galmarini, "Antecedentes sobre la organización del Servicio Meteorológico Argentino y su organización actual," *Revista meteorológica* 1, no. 2 (Apr. 1942): 24–25; Santiago Mauro Comerci, *Cronología de la presencia aérea argentina en la Antártida* (Buenos Aires: Instituto Antártico Argentino, 1995), 3–7.

30. Rudloff, "Ozeanflugwetterdienst," 153–158; "The *Bulletin* Interviews: Professor H. Flohn," *World Meteorological Organization Bulletin*, July 1983: 188–189;

Hermann Flohn, "Meteorologie im Übergang Erfahrungen und Erinnerungen (1931–1991)," *Bonner Meteorologische Abhandlungen* 40 (1992): 15–16, 51.

31. Lewis Pyenson, a major proponent of the thesis that imperial activities failed to influence the basic content of the exact sciences in Europe, also overlooked Franz Linke's lasting ability to inspire interest in tropical meteorology among leading German meteorologists after his 1904–1907 sojourn in Samoa; see *Cultural Imperialism and Exact Sciences: German Expansion Overseas, 1900–1930* (New York: Peter Lang, 1985), ch. 3; cf. Paolo Palladino and Michael Worboys, "Science and Imperialism," *Isis* 84, no. 1 (Mar. 1993): esp. 99–102.

32. Nunn, *Yesterday's Soldiers*, 177–179; Newton, *The Perilous Skies*, ch. 5; quotes come from primary sources cited by Newton.

33. Newton, *The Perilous Skies*, 16–17, 21, 103 [quote], 125–133; Rosalie Schwartz, *Pleasure Island: Tourism and Temptation in Cuba* (Lincoln: University of Nebraska Press, 1999), 65–67.

34. The rise of Pan American Airways and its takeover of SCADTA have attracted enormous attention; see Newton, *The Perilous Skies*, esp. 309–316; Davies, *Airlines of Latin America*, 224–226, 232–236, 628–631; Davies, *A History of the World's Airlines*, ch. 9, esp. table 16; Rinke, *"Der letzte freie Kontinent"*, 2: 724–727, 734–737; Marylin Bender and Selig Altschul, *The Chosen Instrument: Pan Am, Juan Trippe: The Rise and Fall of an American Entrepreneur* (New York: Simon & Schuster, 1982), esp. pt. 2; William A. M. Burden, *The Struggle for Airways in Latin America* (New York: Council on Foreign Relations, 1943), 22–32, 56–59, 115–118, 128.

35. Cushman, "Enclave Vision: Foreign Networks in Peru and the Internationalization of El Niño Research during the 1920s," *History of Meteorology* 1, no. 1 (2004): 67–68.

36. U.S. Hydrographic Office, *Naval Air Pilot: Mexico* (Washington, DC, 1933); A. V. Kidder, "Five Days over the Maya Country," *The Scientific Monthly* 30, no. 3 (Mar. 1930): 193–205; Newton, *The Perilous Sky*, 195, 267–268, 323–324; C. E. Palmer, "Tropical Meteorology," in *Compendium of Meteorology* (Boston: American Meteorological Society, 1951), 864–865 [quote].

37. Burden, *The Struggle for Airways*, 17–18, 29–32, 49–54; Davies, *Airlines of Latin America*, 191–192, 288–289, 293, 383–387, 554. See also Eric Paul Roorda, "The Cult of the Airplane among U. S. Military Men and Dominicans during the U. S. Occupation and the Trujillo Regime," in *Close Encounters of Empire: Writing the Cultural History of U. S.–Latin American Relations*, ed. Gilbert M. Joseph, Catherine C. LeGrande, and Ricardo D. Salvatore (Durham, NC: Duke University Press, 1998), 268–310.

38. Donald M. Pattillo, *Pushing the Envelope: The American Aircraft Industry* (Ann Arbor: University of Michigan Press, 1998), 94–97 [quote], 100; Roger E. Bilstein, *The Enterprise of Flight: The American Aviation and Aerospace Industry*

(Washington, DC: Smithsonian Institution Press, 2001), 56–60; Davies, *A History of the World's Airlines*, 133–136.

39. Burden, *The Struggle for Airways*, 39–44, 47–48, 95–97, map 7; Davies, *Airlines of Latin America*, 403–407, 418–419, 526.

40. Ludwig Weickmann to Galmarini, 12 July 1937; Galmarini to Juan Odermatt, 20 Aug. 1937; "Temario para la Segunda Reunión de la Comisión Regional III en Montevideo, Febrero 1939," 10 Oct. 1938; Präsident der Deutsche Seewarte to Observatorio de Astronomía y Meteorología de Quito, 3 Sept. 1938; all AOAQ-CR. Galmarini, "Nuevos horizontes de la meteorología continental americana," *BAMS* 24, no. 1 (Jan. 1943): 2–4; Otto Schneider, "Relato sobre los trabajos para la intercomparación de los barómetros patrones sudamericanos," *Proceedings of the Eighth American Scientific Congress Held in Washington, May 10–18, 1940* (Washington, DC: Department of State, 1941), 7:331–335; Rudloff, "Ozeanflugwetterdiesnt," 156, 158.

41. Lucien L. Friez to Director del Observatorio, 5 Jan. 1939, AOAQ-CR. Pro-German sentiment among South American meteorologists had been years in the making, see Hoxmark, *La aviación y la meteorología*, 12; Sampaio Ferraz, *Meteorologia brasileira*, 419.

42. See Michael Bravo and Sverker Sörlin, eds., *Narrating the Arctic: A Cultural History of Nordic Scientific Practices* (Canton, MA: Science History Publications, 2002); Trevor H. Levere, *Science and the Canadian Arctic: A Century of Exploration, 1818–1918* (Cambridge: Cambridge University Press, 1993), esp. 362–377.

43. V. Bjerknes, "The Meteorology of the Temperate Zone and the General Atmospheric Circulation," *Monthly Weather Review* 49, no. 1 (Jan. 1921): 2–3; Bjerknes quoted in Robert Marc Friedman, "Constituting the Polar Front," *Isis* 73, no. 3 (Sept. 1981): 361–362. For broader context, see Friedman, *Appropriating the Weather: Vilhelm Bjerknes and the Construction of a Modern Meteorology* (Ithaca, NY: Cornell University Press, 1989); Frederik Nebeker, *Calculating the Weather: Meteorology in the 20th Century* (San Diego, CA: Academic Press, 1995).

44. Exceptions to this include Joel B. Hagen, "Clementsian Ecologists: The Internal Dynamics of a Research School," *Osiris*, 2nd ser., 8 (1993): 178–195; and Margaret W. Rossiter, *The Emergence of Agricultural Science: Justus Liebig and the Americans, 1840–1880* (New Haven, CT: Yale University Press, 1975). See also Gerald L. Geison and Frederic L. Holmes, eds., "Research Schools: Historical Reappraisals," *Osiris*, 2nd ser., 8 (1993); Geison, "Scientific Change, Emerging Specialties, and Research Schools," *History of Science* 19 (1981): 20–40; J. B. Morrell, "The Chemist Breeders: The Research Schools of Liebig and Thomas Thomson," *Ambix* 19 (1972), 1–46.

45. J. Bjerknes and M. A. Giblett, "An Analysis of a Retrograde Depression in the Eastern United States of America," *Monthly Weather Review* 52, no. 11 (Nov.

1924): 521–527; Carl-Gustaf Rossby and Richard Hanson Weightman, "Application of the Polar-Front Theory to a Series of American Weather Maps," *Monthly Weather Review* 54, no. 12 (Dec. 1926): 485–496 [quote p. 496].

46. Horace R. Byers, "Carl-Gustaf Rossby, the Organizer," in *The Atmosphere and the Sea in Motion: Scientific Contributions to the Rossby Memorial Volume*, ed. Bert Bolin (New York: Rockefeller Institute Press, 1959), 56–57; Roscoe R. Braham, Jr. and Thomas F. Malone, *Horace Robert Byers, 1906–1998*, Biographical Memoirs, vol. 79 (Washington, DC: National Academy Press, 2001), 4–6; Sverre Pettersen, *Weathering the Storm: Sverre Petterssen, the D-Day Forecast, and the Rise of Modern Meteorology*, ed. James Rodger Fleming (Boston: American Meteorological Society, 2001), 70–75, 79–81, 317.

47. "The *Bulletin* Interviews: Professor H. Flohn," 189.

48. These trends are also dealt with by Kristine C. Harper, "Boundaries of Research: Civilian Leadership, Military Funding, and the International Network Surrounding the Development of Numerical Weather Prediction in the United States," (Ph.D. diss., Oregon State University, 2003), esp. 62–63, 71, 78–80, 83, 112–113, 145.

49. Arnt Eliassen, *Jacob Aall Bonnevie Bjerknes, 1897–1975*, Biographical Memoirs, vol. 68 (Washington, DC: National Academy Press, 1996), 11–12.

50. Thompson and Seager, *Canada 1922–1939*, esp. 40–49, 53, 76–91, 175, 305.

51. Morley Thomas, *The Beginnings of Canadian Meteorology* (Toronto: ECW Press, 1991), esp. 179, 183, 197–200, 252–253, 255, 258–261; Patterson, "A Century of Canadian Meteorology," 26–28 [quote]; Philip Smith, *It Seems Like Only Yesterday: Air Canada, the First 50 Years* (Toronto: McClelland & Stewart, 1986), 12–14, 18, 20–23; Davies, *A History of the World's Airlines*, 82–87, 209–211.

52. Patterson, "A Century of Canadian Meteorology," 28–29; "Minutes of the Joint Meeting at Toronto of the Royal Meteorological Society and the American Meteorological Society on August 28 and 29, 1939," *QJRMS* 66, supplement (1940): 6–7; Thomas, "The Formation and Early Days of the Canadian Branch of the Royal Meteorological Society," *CMOS Bulletin SCMO* 22, no. 1 (Feb. 1994): 7–13; Petterssen, *Weathering the Storm*, 68; Edward A. Doty to Louis B. Myers, 17 Mar. 1977, box 1:5, UCLA-Bjerknes; Smith, *It Seems Like Only Yesterday*, 29, 35, 54, 56–57, 59, 62; David MacKenzie, *Canada and International Civil Aviation, 1932–1948* (Toronto: University of Toronto Press, 1989), 27–32, 36–38, 54–55.

53. "Minutes of the Joint Meeting at Toronto," 5–6, 8–10.

54. Rossby spent most of his time in Toronto discussing the most important publication of his storied career, "Relation between Variations in the Intensity of the Zonal Circulation of the Atmosphere and the Displacement of the Semi-Permanent Centers of Action," *Journal of Marine Research* 2, no. 1 (1939): 38–55. He submitted arguably his second most important article for publication

in the Toronto proceedings, "Planetary Flow Patterns in the Atmosphere," *QJRMS* 66, supplement (1940): 68–87; for other seminar participants' contributions, see ibid., 57–67, 88–111. Cf. Nebeker, *Calculating the Weather*, 88–90.

55. A fully independent Canadian Meteorological Society was formed on 1 Jan. 1967; "Minutes of the Joint Meeting at Toronto," 14–15; cf. Thomas, "The Formation and Early Days of the Canadian Branch."

56. "International Union of Geodesy and Geophysics: Meteorological Association, Program for the Washington Assembly," *BAMS* 20, no. 6 (June 1939): 274–275; J. Bjerknes, P. Mildner, E. Palmén, and L. Weickmann, *Synoptisch-aerologische Untersuchung der Wetterlage während der internationalen Tage vom 13. bis 18. Dexember 1937* (Leipzig: Veröffentlichungen des Geophysikalischen Instituts der Universität Leipzig, 1939), esp. 3–6; Eliassen, *Jacob Aall Bonnevie Bjerknes*, 11; "Minutes of the Joint Meeting at Toronto," 11; Nebeker, *Calculating the Weather*, 104–106. Cf. Deborah Day, "Bergen West: Or, How Four Scandinavian Geophysicists Found a Home in the New World," *Historisch-Meereskundliches Jahrbuch* 6 (1999): 69–82.

57. Mrs. Robert G. Sproul to H. U. Sverdrup, 20 Mar. 1936; Sverdrup to Robert G. Sproul, 11 Apr. 1936; Csüdriu Sverdrup to Mrs. T. Wayland Vaughan, 6 May 1936; 1:8, SIO-Sverdrup. J. C. Harper to Sproul, 8 Jan. 1938; Sproul to Harper, 19 Jan. 1938, UCLA-SIO; Friedman, "Contexts for Constructing an Ocean Science: The Career of Harald Ulrik Sverdrup (1888–1957)," in *Oceanographic History: The Pacific and Beyond*, ed. Keith R. Benson and Philip F. Rehbock (Seattle: University of Washington Press, 2002), 21–23; Eric L. Mills, "The Oceanography of the Pacific: George F. McEwen, H. U. Sverdrup and the Origin of Physical Oceanography on the West Coast of North America," *Annals of Science* 48, no. 3 (May 1991): 261–262; Tor Bergeron, "The Young Carl-Gustaf Rossby," in *The Atmosphere and the Sea in Motion*, 53–54.

58. Burton M. Varney to Vern O. Knudson, 16 Jan. 1940; Rossby to Sproul, 7 Mar. 1940; Sverdrup to Sproul, 19 Mar. 1940; Sproul to Bjerknes, 22 Apr. 1940; Bjerknes to Sproul, 2 May 1940, 5 Aug. 1940; 84:1, UCLA-PhysMet.

59. See Kevin Starr, *The Dream Endures: California Enters the 1930s* (New York: Oxford University Press, 1997), ch. 3; Judith R. Goodstein, *Millikan's School: A History of the California Institute of Technology* (New York: W. W. Norton, 1991); J. M. Lewis, "Cal Tech's Program in Meteorology: 1933–1948," *BAMS* 75, no. 1 (1994): 69–82; Roger W. Lotchin, *Fortress California, 1910–1961: From Warfare to Welfare* (New York: Oxford University Press, 1992), ch. 4.

60. Rossby to Sproul, 3 May 1940; G. M. Richards to Regents of UCLA, 20 June 1940; Joseph Kaplan to Deming G. Maclise, 18 Dec. 1941; Reichelderfer to Bjerknes, 5 Nov. 1941, 12 Dec. 1941; 84:1, 108:14, UCLA-PhysMet.

61. Bjerknes to Sproul, 2 May 1940, 11 May 1940; Holmboe to Bjerknes, 24 June 1940; Holmboe to Kaplan, 10 July 1940; Sproul to Holmboe, 30 or 31 [?] Aug.

1940; 84:1, UCLA-PhysMet. For a contrasting viewpoint, see Sverdrup to Sproul, 1 May 1940, 1:16, SIO-Sverdrup.

62. Sverdrup to Sproul, 30 Apr. 1942; George Turner to Sproul, 21 May 1942; 139:9 UCLA-SIO. See also Naomi Oreskes and Ronald Rainger, "Science and Security before the Atomic Bomb: The Loyalty Case of Harald U. Sverdrup," *Studies in History and Philosophy of Modern Physics* 31, no. 3 (2000): 309–369.

63. Andrew Thomson served as chief organizer of meteorological training for the British Commonwealth Air Training Plan. Bjerknes to Sproul, 5 Aug. 1940; Rossby to Bjerknes, 13 Aug. 1940; Reichelderfer to Bjerknes, 13 Aug. 1940; Kaplan to Sproul, 24 Feb. 1942; "Announcement of Program in Meteorology, 1942," Mar. 1942; 84:1, 136:2 UCLA-PhysMet. "Scripps Institution of Oceanography of the University of California," Report to the Regents of the University of California, [1945], 87:18, UCLA-SIO. Bjerknes to Sproul, 1 Feb. 1946, 223:19, UCLA-Met. "District Forecast Center at Los Angeles," *BAMS* 21, no. 7 (Sept. 1940): 309; Thomas, "The Formation and Early Days of the Canadian Branch," 9–10; Thomas, "Andrew Thomson," Canadian Meteorological and Oceanographic Society, http://www.cmos.ca/prizebios.html (accessed 19 Oct. 2005).

64. These acts helped rehabilitate Ludwig Weickmann's reputation after the war. Weickmann to Carl Störmer, 31 Oct. 1946; Störmer, Armed Forces affidavit, 27 Dec. 1946; 1:5, UCLA-Bjerknes. Hermann Flohn, "Ludwig Weickmann zum Gedächtnis," *Beiträge zur Physik der Atmosphäre* 35, no. 3/4 (1962): 141–144.

65. Sverdrup's emphasis; Sverdrup to Gorczynski, 19 Sept. 1941, 1:19, SIO-Sverdrup. Sverdrup to Sproul, 22 Nov. 1939; J. H. Corley to Maclise, 16 May 1941; 112:1, UCLA-SIO. Ladislas Gorczynski, *Breve reseña sobre observaciones de radiación solar y aparatos termoeléctricos registradores: Importancia de su estudio en la República Mexicana* (Tacubaya, Mexico: Imprenta de la Dirección de Estudios Geográficos y Climatológicos, 1926), 1–6, 14, 18; Carlos Theye, "Radiación solar en la Habana," *Boletín del Observatorio Nacional* (Havana) 22, no. 2 (1926): 19–32; Theye, "Estudios comparativos de radiación solar (1)," *Boletín del Observatorio Nacional* 24, no. 1 (1928): 51–74; Gorczynski to Luis G. Tufiño, 20 May 1928, AOAQ-CR.

66. Canada did not generally participate in "Pan-American" events; *Proceedings of the Eighth American Scientific Congress Held in Washington, May 10–18, 1940* (Washington, DC: Department of State, 1941), 1:9–10, 13, 15–16, 24, 31, 35.

67. *Proceedings of the Eighth American Scientific Congress*, 1:153–154; José C. Gómez, "Desarrollo del Servicio Meteorológico de México" (ibid., 7:433–436); C. F. Brooks, "The Joint Meeting of the Eighth American Scientific Congress and the American Meteorological Society, Washington, D.C., May 14, 1940," *BAMS* 22, no. 2 (Feb. 1941): 77–81.

68. Brooks, "A Supplement to This Issue of the Bulletin," *BAMS* 23, no. 10 (Dec. 1942): 420; Edward H. Bowie, "Latin American Section," *BAMS* 24, no. 1 (Jan.

1943): 1–2, emphasis added; Galmarini, "Nuevos horizontes de la meteorología continental americana," *BAMS* 24, no. 1 (Jan. 1943): 2–4; Morandi, "Nuestro programa," *Revista meteorológica* 1, no. 1 (Jan. 1942): 1–6; "Creación de la Junta Nacional de Meteorología de Uruguay" (ibid., 19); Galmarini to Odermatt, 31 Mar. 1944, AOAQ-IMO Region III correspondence.

69. Nunn, *Yesterday's Soldiers*, 177–178, 194, 203–204, 207, 210–211, 216–217, 219; Davies, *Airlines of Latin America since 1919*, 230–232, 321; Jorge Salvador Lara, *Breve historia contemporanea del Ecuador* (Mexico City: Fondo de Cultura Económica, 2000), 462–464; Alberto Fernández Prada E., *La aviación en el Perú, 1751–1942* (Lima: Editorial CIMP, 1966), 1:627–636; Frank D. McCann, Jr., "Aviation Diplomacy: The United States and Brazil, 1939–1941," *Inter-American Economic Affairs* 21, no. 4 (Spring 1968): 48.

70. Burden, *The Struggle for Airways*, xxi–xxii [quote]. For a scathing examination of U.S. anti-German motives and practices, see Max Paul Friedman, *Nazis and Good Neighbors: The United States Campaign against the Germans of Latin America in World War II* (Cambridge: Cambridge University Press, 2003); and Stephen James Randall "Colombia, the United States, and Interamerican Aviation Rivalry, 1927–1940," *Journal of Interamerican Studies and World Affairs* 14, no. 3 (Aug. 1972): 297–324.

71. Davies, *Airlines of Latin America since 1919*, 236–240, 273–274, 297–298, 324–326, 407–417, 527, 605–623; Burden, *The Struggle for Airways*, 45, 51, 69–77; Claus Wollheim, *Die Geschichte der deutsch-kolumbianischen Luftverkehrsgesellschaft SCADTA unter besonderer Berücksichtigung ihres Postwesens* (Zurich, 1978), 25–26; Boy, *Una historia con alas*, 239–241, 246–249; Arias, *Otro cóndor sobre los Andes*, 100–103; Friedman, *Nazis and Good Neighbors*, 2; Villa, *Alas de Bolivia*, 1:165–167; McCann, "Aviation Diplomacy," 35–50; John F. Fuller, *Thor's Legions: Weather Support to the U.S. Air Force and Army, 1937–1987* (Boston: American Meteorological Society, 1990), 64; John Wirth, *The Politics of Brazilian Development, 1930–1954* (Stanford: Stanford University Press, 1970), 1, 28–39, 51–68, 79–80, 116–118.

72. Willington Paredes Ramírez, *Historia social de Salinas* (Guayaquil: Archivo Histórico del Guayas, 2004), 274–283; Jorge Villacrés Moscoso, *Historia diplomática de la República del Ecuador* (Guayaquil: Editorial Guayaquil, 1978), 46–52.

73. Fuller, *Thor's Legions*, 64; Leo Alpert, *Pilot Interviews: Composite of 7,500 Trips through the Tropical Front*, Air Force Weather Technical Document AWS TR 105-52 (ATI-65123), Nov. 1944; Alpert, "The Intertropical Convergence Zone of the Eastern Pacific Region," *BAMS* 26 (1946): 426–432, vol. 27 (1946): 15–29, 62–66; Alpert, "Notes on the Weather and Climate of Seymour Island, Galapagos Archipelago," *BAMS* 27 (1946): 200–209; Alpert, "Weather over the Tropical Eastern Pacific Ocean, 7 and 8 March, 1943," *BAMS* 27 (1946): 384–398.

74. José C. Gómez to Jefe del Servicio Meteorológico del Ecuador, 24 Nov. 1942; Odermatt to Col. W. E. Shipp, 19 Nov. 1943; AOAQ-CR. Odermatt to Weightman, 21 May 1945, AOAQ-USWB; Odermatt to Petterssen, 8 Oct. 1948, AOAQ-IMO [quote]; see also Herbert Real and Leo Alpert, *The El Nino Event of 1943: A Survey of Its History and of Observations and Methods of Computation Available for Its Analysis* (Boulder, CO: Cooperative Institute for Research in the Environmental Sciences, 1981).

75. David Green, *The Containment of Latin America: A History of the Myths and Realities of the Good Neighbor Policy* (Chicago: Quadrangle Books, 1971), 47–48; Frederick B. Pike, *FDR's Good Neighbor Policy: Sixty Years of Generally Gentle Chaos* (Austin: University of Texas Press, 1995), 251–254.

76. "Application for Special Leave of Absence: Jacob Bjerknes," 9 Dec. 1940; Bjerknes to E. R. Headrick, 22 Dec. 1940; 84:1, UCLA-PhysMet. Odermatt to Ministro de Educación Pública, 8 Apr. 1941, AOAQ-USWB; Julio Lamarthée, "La próxima reunion meteorológica en Washington," *Revista meteorológica* 1, no. 3 (July 1942): 67–70; Weightman, "Meteorological Training for Latin-American Students," *BAMS* 25, no. 4 (Dec. 1944): 435. On Weightman's early ties to the Bergen school, see Weightman, "Some Observations on the Cyclonic Precipitation of February 22–23, 1925, in the Central and Eastern United States," *Monthly Weather Review* 53, no. 9 (Sept. 1925): 379–384; Rossby and Weightman, "Application of the Polar-Front Theory to a Series of American Weather Maps."

77. Rossby to Benton, "Cooperation with Latin-American Meteorologists" [n.d.]; Rossby to Sproul, 22 Nov. 1941; Kaplan to Sproul, 2 Sept. 1941, 7 Oct. 1941, 20 Oct. 1941; 108:14, UCLA-PhysMet.

78. B. C. Haynes, *Meteorología para pilotos de aviones*, 2nd ed. (Washington, DC: Government Printing Office, 1943), esp. iii–iv, 3–4, 58, 65–68, 115, 135–139, 251–254, 275–276; Abelardo Montalvoto to Director del Observatorio, 16 Feb. 1943, AOAQ-CR.

79. Weightman, "Meteorological Training for Latin-American Students," 435–436; Charles F. Sarle to Kaplan, 14 Aug. 1942; Sarle, "Memorandum to the Universities Teaching Professional Meteorology," 23 July 1943; Kaplan to Sproul, 4 Aug. 1943; Sarle to Gordon S. Watkins, 23 May 1944; 136:2, 156:9, 175:1, UCLA-PhysMet.

80. See Starr, *The Dream Endures*, 92–94, 103–105, 109, 111–112; Starr, *Inventing the Dream: California through the Progressive Era* (New York: Oxford University Press, 1985), ch. 2; Lotchin, *Fortress California*, ch. 2; Gray Brechin. *Imperial San Francisco: Urban Power, Earthly Ruin* (Berkeley and Los Angeles: University of California Press, 1999), ch. 5.

81. Russell H. Fitzgibbon to Editor of *En Guardia*, 17 Oct. 1942; George F. Taylor to James H. Corley, 6 July 1944; Bjerknes to Sproul, 3 July 1944; "Graduation

of Students in Class Seven of the War Training Program in Meteorology," 3 June 1944; Sarle to Taylor, 11 Oct. 1944; 136:2, 175:1, UCLA-PhysMet. Taylor to Bjerknes, 29 June 1945; Department of Commerce Weather Bureau, "Contract for University Training in Meteorology," 29 Oct. 1945; 193:12, UCLA-Met.

82. Weightman, "Meteorological Training for Latin-American Students," 435–436; "Memorandum: Alumnos Ecuatorianos que asistieron al curso de meteorología . . . en Medellín bajo los auspicios del U.S. Weather Bureau," 10 Jan. 1944, AOAQ-Informes. Odermatt to Reichelderfer, 7 July 1948; Vicente L. Gómez to Mission Chief for the Institute of Interamerican Affairs, 16 June 1952; AOAQ-USWB. Headrick to M. de Freyre y Santander, 17 Mar. 1942, 136:2, UCLA-PhysMet; Ernesto More, *Reportajes con radar* (Lima: Librería e Imprenta Minerva, 1960), 405–411 [quote].

83. See Arturo Escobar, *Encountering Development: The Making and Unmaking of the Third World* (Princeton, NJ: Princeton University Press, 1995).

84. A. Serra and L. Ratisbonna, *As masas de ar da America do Sul* (Rio de Janeiro: Serviço de Meteorologia, 1942); Ricardo Toscano, *Meteorología descriptiva y dinámica, climatología: Para el curso de la Escuela de Ingeniería de la Universidad Nacional Autónoma de México* (Mexico City: Imprenta Universitaria, 1950), esp. pt. 3; Odermatt to Reichelderfer, 13 Feb. 1947, AOAQ-USWB.

85. Eduardo Palomo to Director del Observatorio, 4 May 1942; Abelardo Montalvo to Director del Observatorio, 24 Apr. 1942, AOAQ-CR. "Exposición del Director del Observatorio Astronómico de Quito al Señor Ministro de Agricultura," Jan. 1944, AOAQ-Informes. Weightman to Odermatt, 6 Apr. 1945; Odermatt to Weightman, 21 May 1945; Reichelderfer to Odermatt, 28 July 1944, 27 Nov. 1944; AOAQ-USWB.

86. Odermatt to Luis Eduardo Mena, 8 July 1943; Mena to Odermatt, 19 July 1943; Reichelderfer to Odermatt, 24 Sept. 1943, AOAQ-CR. Odermatt to Reichelderfer, 4 May 1944, 4 May 1945; Reichelderfer to Odermatt, 8 June 1944, 3 Apr. 1946; Odermatt to Weightman, 21 May 1945, AOAQ-USWB.

87. Don M. Coerver and Linda B. Hall, *Tangled Destinies: Latin America and the United States* (Albuquerque: University of New Mexico Press, 1999), 110–130; Walter LaFeber, *America, Russia, and the Cold War, 1945–2000*, 9th ed. (Boston: McGraw-Hill, 2002), 23–24, 72–73, 124–125, 165, 214–219; Clark A. Miller, "Scientific Internationalism in American Foreign Policy: The Case of Meteorology, 1947–1958," in *Changing the Atmosphere: Expert Knowledge and Environmental Governance*, ed. Clark A. Miller and Paul N. Edwards (Cambridge, MA: MIT Press, 2001), esp. 191–198; Cushman, "Choosing between Centers of Action: Instrument Buoys, El Niño, and Scientific Internationalism in the Pacific, 1957–1982," in *The Machine in Neptune's Garden: Historical Perspectives on Technology and the Marine Environment*, ed. Helen M. Rozwadowski and David K. van Keuren (Sagamore Beach, MA: Science History Publications, 2004), esp. 141–144.

88. "Organización del Servicio Meteorológico de Chile y plan de desarrollo del mismo," *Revista meteorológica* 1, no. 3 (July 1942): 24–33; Julio Bustos Navarrete, "Las sequias y los años lluviosos en Chile durante los últimos cuatro siglos," *Proceedings of the Eighth American Scientific Congress*, 7:395–397; "Las lluvias en la región central de Chile hasta el año 1824" (ibid., 397–399); "Las investigaciones realizadas en Chile por el Observatorio del Salto, sobre la radiación solar y el tiempo" (ibid., 399–402); "La acción de las variaciones de la radiación solar en la circulación atmosférica del Pacífico" (ibid., 402–403); "Relación entre el regimen meteorológico de la Indochina y Chile" (ibid., 403–404). Cf. Sampaio Ferraz, "Suggestions for the Explanation of Probable Connections between Solar Activity and Rainfall Variation in Southeastern Brazil" (ibid., 373–376); Sampaio Ferraz, *Meteorologia brasileira*, 178, 242, 303–307; Cushman, "Walker, Gilbert Thomas," in *New Dictionary of Scientific Biography*.

89. J. Boerema to E. C. Heal, 22 May 1933; Georg Petersen to Director del Observatorio, 25 March 1938; AOAQ-CR. Odermatt, "El clima de Quito en el año 1941," AOAQ-Informes. Odermatt, "La influencia del sol sobre el regimen de las lluvias en el Ecuador," *Proceedings of the Eighth American Scientific Congress*, 7:413–417; Aquiles R. Pérez, "Las precipitaciones anormales en el territorio de la República del Ecuador" (ibid., 421–422).

90. Nebeker, *Calculating the Weather*, 94–99; Cushman, "Choosing between Centers of Action."

91. Weightman, "Meteorological Training for Latin-American Students," 435–436; "Clarence E. Palmer, Geophysics and Planetary Physics: Los Angeles, 1911–1973," *University of California: In Memoriam*, Mar. 1976, available at http://ark.cdlib.org/ark:/13030/hb9k4009c7; Riehl, *Tropical Meteorology*, v; Robert G. Stone, "On the Mean Circulation of the Atmosphere over the Caribbean," *BAMS* 23, no. 1 (Jan. 1942): 4–16; Patrick J. Fitzpatrick, *Natural Disasters: Hurricanes: A Reference Handbook* (Santa Barbara, CA: ABC-CLIO, 1999), 119–120.

92. Riehl, *Waves in the Easterlies and the Polar Front in the Tropics*, University of Chicago, Department of Meteorology, Miscellaneous Reports No. 17 (Chicago: University of Chicago Press, 1945), passim; Palmer, "Tropical Meteorology," 869–870; Riehl, *Tropical Meteorology*, 235–243; Gordon E. Dunn and Banner I. Miller, *Atlantic Hurricanes* (Baton Rouge: Louisiana State University Press, 1964), esp. 27–28, 37.

93. Palmer, "Tropical Meteorology," 862–863, 867, 871–872.

94. W. H. B. Rudloff, *Meteorology in Peru*, United Nations Technical Assistance Programme, Report No. TAA/PER/3 (New York, 1959), 1–6, 36–37 [quote]; Odermatt to Weightman, 21 May 1945, AOAQ-USWB.

95. Aníbal Pinto and Oswaldo Sunkell, "Latin American Economists in the United States," *Economic Development and Cultural Change* 15, no. 1 (Oct. 1966): 79–86.

96. "The *Bulletin* Interviews: Professor H. Flohn," 191–192; Rudloff, "Ozeanflug-wetterdienst," 154; Flohn, "Nachrufe: Prof. Dr. Werner Schwerdtfeger," *Mitteilungen der Meteorogische Gesellschaft* 2 (1985): 56–57.

97. Palmer, "Tropical Meteorology," 871–872; Fitzpatrick, *Natural Disasters: Hurricanes*, 119–120; "Clarence E. Palmer"; Cushman, "Viñes, Benito."

98. Jonathan Harwood, "National Styles in Science: Genetics in Germany and the United States between the World Wars," *Isis* 78, no. 3 (Sept. 1987): 390–414; Marga Vicedo, "Scientific Styles: Toward Some Common Ground in the History, Philosophy, and Sociology of Science," *Perspectives on Science* 3, no. 2 (Summer 1995): 231–254.

99. Kwame Nkrumah (1965), quoted in Young, *Postcolonialism*, 44; Wolfgang Sachs, "Introduction" to *The Development Dictionary: A Guide to Knowledge as Power* (London: Zed Books, 1992), 1–5; Ramachandra Guha, "The Age of Ecological Innocence," in *Environmentalism: A Global History* (New York: Longman, 2000), 63–68.

CHAPTER 8

Global Climate Change and Human Agency

Inadvertent Influence and "Archimedean" Interventions

JAMES RODGER FLEMING

Wallace Broecker has referred to the global climate sytem
as a "massive staggering beast." To emphasize rapid
climate changes, Richard Alley added the adjective
"drunken." When I introduced Richard's lecture at Colby
last year (2005), I noted that it was not wise to anger
such a beast, yet this is just what we were doing with
anthropogenic emissions, which were, in effect, our sticks
to prod the beast. Since then (2006), I have become
interested in longer sticks, Archimedean levers to "tip"
the climate system in supposedly favorable directions
though geoengineering. But what would be the
consequences of such intervention?

TIPPING POINTS: PHYSICAL AND SOCIAL

At the American Geophysical Union meeting in San Francisco in December
2005, climate scientist James Hansen of NASA warned that the Earth's climate was nearing an unprecedented "tipping point"—a point of no return

that could only be avoided if the "growth of greenhouse gas emissions is slowed" in the next two decades:

> The Earth's climate is nearing, but has not passed, a tipping point beyond which it will be impossible to avoid climate change with far-ranging undesirable consequences. These include not only the loss of the Arctic as we know it, with all that implies for wildlife and indigenous peoples, but losses on a much vaster scale due to rising seas . . . . This grim scenario can be halted if the growth of greenhouse gas emissions is slowed in the first quarter of this century.[1]

Hansen's brief statement, widely distributed by the press, clearly struck a cultural nerve. It acknowledged undesirable inadvertent human influence on the climate system and pointed to a possible remedy. In the interest of impact, however, it avoided complexities. For example, it is highly unlikely that merely *slowing the growth* of emissions would be very effective. Missing were the net positive results of a true behavioral "tipping point" beyond which humanity decided to live with only clean energy. Another "tipping point" also went unmentioned: the growing international agenda to intervene purposefully in the global climate through geoengineering.

The April 3, 2006 cover of *Time* magazine captured the essence of the apocalyptic mood on climate change, advising its readers to "Be worried, be *very* worried," since the "climate is crashing, and global warming is to blame." Focusing on what can be done politically and economically to avoid dangerous climate change, the May issue of *Vanity Fair* reported that the first imperative is to "punch through the massive denial and resistance" that still exists in the United States.[2]

If indeed a "tipping point" in the global climate system is imminent, can changes in collective lifestyle avert an unprecedented and potentially catastrophic disaster? Will mitigation of human influence do the job? Or must we settle for adaptation? What level of mitigation? Who is responsible? Who pays and how much? To what ultimate effect? There is another more ominous option. Should technical elites "fix" the climate and attempt to control nature through geoengineering? Are there other options? We may be witnessing the process by which we convince ourselves that the Earth's climate needs to be "fixed," thus passing a sociological "tipping point" that would favor purposeful intervention, through mitigation of human actions, adaption to adverse consequences, or far more drastic intervention in the climate system itself.[3]

FIGURE 1   Give me a place to stand and I will move the Earth"—Archimedes. Engraving from *Mechanics Magazine* (London, 1824).

In recent decades the rise of climate dynamics, in the form of computer climate models and global monitoring systems, have led some to conclude that future climate states can be predicted and that climate (or at least human impact on the climate) might even be managed or controlled. In other words, in an Archimedean sense (Figure 1), climate might have a physical "tipping point," and humans may now have levers long enough to move it. But where will the Earth roll when you tip it (Figure 2)?

A ROLE FOR HISTORY

Global climate change is an international, intergenerational, and interdisciplinary topic. Yet most scholarship focuses on the West, the present, and the science or economics of the issues. Historians of the topic are fewer than few, as are interesting ethicists and other humanistic scholars.[4] In this essay, I hope to provide a partial corrective by bringing the history of human influence on the climate—both inadvertent and intentional—to bear on questions

FIGURE 2   But where will the Earth roll when you tip it?

of ethics and public policy. In doing so I will be setting out a broad research agenda linking the three disciplines.

What have people experienced, learned, feared about climate change in the past? How have they proposed to intervene? By what paths have we reached the current state of climate apprehension? Since we are literally immersed in the climate system, how are priviledged positions developed and maintained? How can a historical analysis of the modes of climate intervention, whether proposed or practiced, inadvertent or intentional, shed light on the pressing issues of our day and perhaps help us evaluate or at least envision future possibilities.[5]

Anthropogenic climate change is not at all a new issue. In eras other than our own, the climate has been perceived as amenable to human impact or intervention. In fact, the question of human agency, the ways in which humanity had purposefully changed the Earth (including the climate) from its hypothetical pristine condition, was one of three perennial questions about nature and culture posed by the intellectual ancestors of the Western tradition.[6] In the eighteenth century, Enlightenment thinkers argued that human settlement had caused a gradual warming of the European continent. Settlers in the New World engaged in self-conscious, if ultimately ineffective efforts to modify and "improve" the climate through clearing the forests and cultivating the lands. Scientists in the nineteenth century, seeking a natural explanation for climate change, based their arguments on universal or terrestrial physics and largely ignored theories of human influence.

At the turn of the twentieth century, the climatic effects of industrial emissions—especially the rising use of coal—came under scrutiny. Some speculated that anthropogenic warming could possibly have a long-term beneficial effect on climate by staving off the anticipated return of an ice age. Most scientists, however, thought that increasing carbon dioxide levels in the atmosphere would have no effect at all. Again, most theories of climate change were rooted in the study of natural phenomena.

In 1938 G.S. Callendar, using new data on global temperatures, fuel consumption, and infrared absorption, revived the theory that global climate change could be caused by anthropogenic increases in the concentration of atmospheric carbon dioxide. Since then, the so-called "Callendar effect" has become the dominant theory of human agency, and, some believe, the dominant forcing factor overall in global climate change. In recent decades the rise of climate dynamics, in the form of computer climate models and global monitoring systems, have led some to conclude that future climate states can be predicted and that climate (or at least human impact on the climate) might even be managed or controlled.

With rising awareness of potential human damage to the climate system, the term "inadvertent climate change," widely used before about 1970, has now become outdated. We are fully aware, so goes the argument, of our collective impact on the planet, since the Rio conference in 1992 or at least the IPCC third assessment report in 2001. Nobel Laureate Paul Crutzen calls this the "Anthropocene era" of anthropogenic CO_2 and methane, dating it to the 18th century. Environmentalists might label it the "anthro-obscene" era of human influence on climate. Hansen and, recently, British Prime Minister Tony Blair give us 7 years to get our act together before the next round of the post-Kyoto negotiations.

But lest we neglect the past, this essay examines how people in different eras have understood the human influence on climate, how they have intervened, and the value of studying the history of climate modifications. In the space allotted, I cannot undertake a detailed analysis of world history and the ethical and public policy implications of climate engineering. That is my ongoing project.[7]

THE ENLIGHTENMENT NEXUS

Ancient and medieval theories of climate largely centered on questions of direct influence on individuals and indirect influence on the character and nature of human culture. Climates were seen as largely static, a function of latitude and location, and determinism largely overshadowed any notions of human agency. Theophrastus, a student of Aristotle, wrote of local changes of climate caused by human agency, specifically agricultural activities. He observed that draining wetlands removed the moderating effects of water and led to greater extremes of cold, while clearing woodlands for agriculture exposed the land to the Sun and resulted in a warmer climate.[8] The idea that climate influenced culture and human well-being is of course, an ancient one, dating from the Hippocratic corpus and Aristotle's *Politics* through a long series of medieval and early modern climate determinists including Albertus Magnus, Jean Bodin, and John Arbuthnot. [9]

Modern European thought linking climate *change* (the temporal dimension) and culture can be traced to the diplomat, historian, and critic Jean-Baptiste Du Bos (1670–1742), a leading intellectual of his day, member (later perpetual secretary) of the French Academy, and author of the two-volume *Réflexions critiques sur la poësie et sur la peinture* (1719), a book on creativity in relation to geographic factors, especially climate. For Du Bos, the emergence of genius was due primarily to "les causes physiques" (the nature of the air, land, soil, and especially climate), with "les causes morales" (education, cultivation, governance) playing a secondary role. Du Bos argued that artistic genius flourished only in countries with suitable climates (always between twenty-five and fifty-two degrees north); that changes in climate must have occurred to account for the rise and decline of the creative spirit in particular nations in particular eras; and that human agency (clearing and cultivation) had caused the climate of Europe and the Mediterranean area to warm gradually since antiquity, resulting in a decline of creative genius in certain nations. His theory implied that the ongoing deforestation and increased cultivation of North America would result in a rapid change in climate (and culture).

The environmental determinist with the broadest influence from this era was undoubtedly Charles de Secondat Baron de Montesquieu (1689–1755),

228

who was sponsored by Du Bos for membership in the French Academy. While he did not write about climate change per se, Montesquieu agreed with Du Bos that Europeans were at great risk when they *changed climates*, for example, in moving to colonies in Africa, Asia, or the New World. As experience showed, traveling to new lands was risky enough, but settling in new climes would most probably be fatal. According to Montesquieu, the good life was one lived from cradle to grave within the same country, preferably the same region; individuals should stay where they are.

Although Montesquieu wrote that climate was the first of all the empires, it was not the only one. Human ingenuity and effort in areas such as education, government, medicine, and agriculture could overcome the negative influences of climate. Hume asserted that the human response to climate change is of paramount importance: "Man is not simply subject to the necessity of nature; he can and should shape his own destiny as a free agent, and bring about his destined and proper future."[10] Thus moral causes meant much more to him than physical causes, reversing the emphasis of Du Bos.

David Hume (1711–1776) followed Du Bos explicitly on the issue of climate change and human agency when he speculated that by their collective activities, humans had contributed to historical changes in Europe and to much faster changes in America. In his essay "Of the Populousness of Ancient Nations," Hume argued, from evidence found in ancient writings, that the climate of Europe had been colder in ancient times. Hume believed that moderation of the climate had been caused by human settlers who gradually cleared the forest and cultivated the land:

> Allowing, therefore, this remark [of Du Bos] to be just, that Europe is become warmer than formerly; how can we account for it? Plainly, by no other method, than by supposing that the land is at present much better cultivated, and that the woods are cleared, which formerly threw a shade upon the earth, and kept the rays of the sun from penetrating to it.

Hume thought that similar but much more rapid changes were occurring in the Americas as the forests were cleared. "Our northern colonies in America become more temperate, in proportion as the woods are felled."[11]

The ideas of Du Bos, Montesquieu, and Hume dominated the discussion of climate in the second half of the eighteenth century. Du Bos developed an environmental theory of the rise and fall of creative eras. Montesquieu was more interested in the ability of people to govern, even in inhospitable climates, while Hume speculated directly on climate change and human agency in Europe and the Americas. Their arguments appealed directly to European cultural sensibilities and prejudices; the authority of their positions residing in their considerable literary skills and the lack of other evidence to prove them wrong. Collectively they generated a powerful vision of the climates of Europe and America shaping the course of empire and the arts; the concerted efforts of innumerable individuals in turn shaping the climate itself.

HUMAN AGENCY, EUROPEAN DISDAIN, AND THE EARLY AMERICAN CLIMATE

American colonists were confused and confounded by the cold winters and harsh storms they encountered. The New World was a place of considerable hardship for them, and its climate was an object of disdain for many European elites.[12] Flourishing in a rigorous environment and convincing others that the North American continent was not a frozen, primitive, or degenerate wasteland became a crucial element in American apologetics. The notion, held by many settlers and patriots, that such a harsh climate could be improved by human activity—if it is too cold and damp, they could make it better by clearing the forests, draining the marshes, and cultivating the soil—was a major issue in early America and remained so through the early decades of the nineteenth century. If the climate could truly be transformed, the implications were enormous, involving the health, well-being, and prosperity of all.

Thomas Jefferson's *Notes on Virginia* (1787), essentially a patriotic defense of the natural phenomena of the New World, presented an apology for the harsh climate and suggested that it was being improved by settlement. His agenda included a multi-faceted quest for natural historical proof that the American environment was not degenerate, but was either superior to that of Europe or was undergoing rapid improvement by the concerted agency of free citizens, toiling and thriving under its beneficent skies:

A change in our climate . . . is taking place very sensibly. Both heats and colds are become much more moderate within the memory even of the middle-aged. Snows are less frequent and less deep . . . The elderly inform me, the earth used to be covered with snow about

230

three months in every year. The rivers, which then seldom failed to freeze over in the course of the winter, scarcely ever do so now.

Jefferson reasoned that as the forests were cut down, open fields were better able to absorb and retain heat. He speculated that as the settlements of Virginia progressed inland from the seacoast, the sun was heating the cleared and cultivated land and the sea breezes were penetrating farther inland than ever before. The net result would be a temperate climate and clear atmosphere that would serve, in the words of his colleague Hugh Williamson, as "a proper nursery of genius, learning, industry and the liberal arts."[13]

Although they accepted in principle the long list of assumptions about climate change, human agency, and health (see Table 1), American natural philosophers thought that many years of regular and comparative observations would be necessary to settle the issue. Jefferson was a strong advocate for a national meteorological system and encouraged the federal government to supply observers in each county of each state with accurate instruments. In 1824, he wrote to Lewis Beck, "Measurements of the American climate should begin immediately, before the climate has changed too drastically. These measurements should be repeated . . . once or twice in a century, to show the effect of clearing and culture towards the changes of climate."[14]

TABLE 1. Jeffersonian ideas about climate change and human agency

1. Cultures are determined or at least strongly shaped by climate.

2. The climate of Europe has moderated since ancient times.

3. This was caused by the gradual clearing of the forests and by cultivation.

4. The American climate is undergoing rapid and dramatic changes, perhaps ten times more rapidly than Europe.

5. These changes too have been initiated by human agency.

6. The amelioration of the American climate will make it more fit for European-type civilization and less suitable for Native cultures.

7. An "improved" climate will be warmer, less extreme, drier, and healthier.

8. These changes are prologue to much greater changes in both the former colonies and soon, westward across the continent.

9. Widespread and comparable measurements must begin immediately, before the climate has changed too drastically, and must be repeated regularly.

SYSTEMATIC DATA COLLECTION

Beginning in earnest in the second decade of the nineteenth century, physicians natural philosophers, and state and federal agencies, including the General Land Office the Army Medical Department, and the state of New York collected, tabulated, charted, mapped, and otherwise analyzed their weather observations. The process of monitoring the weather (and ultimately the climate) at diverse locations across the country profoundly changed the tenor of the discourse on climate change and established rudimentary foundations for the science of climatology.[15]

With the rise of systematic data collection (including earlier efforts in Europe), the venerable belief in direct influence on climate change and its corollary, immediate human agency, was overshadowed by explanations favoring climate stability and physical causation. In 1843, Dr. Samuel Forry conducted an analysis of weather and health data gathered by the US Army Medical Department at over sixty locations dating back to 1814. Forry concluded that: (1) climates are stable and no accurate thermometric observations warrant the conclusion of climatic change, (2) climates are susceptible of melioration by the changes wrought by the labors of man, but (3) these effects are much less influential than physical geography, including oceans, lakes, mountains, and the dimensions of continents.[16]

The army continued its system of taking meteorological measurements at its posts across the country until 1874, primarily in support of the health of the troops, but also motivated by an attempt to document potential changes in the climate.[17] By then, a complete shift had occurred in climate discourse, now based on empirical evidence and statistical analysis. Cleveland Abbe wrote in 1889, "The true problem for the climatologist to settle during the present century is not whether the climate has lately changed, but what our present climate is, what its well-defined features are, and how they can be most clearly expressed in numbers."

Regional folk theories such as "Rain Follows the Plow,"[18] a promotional campaign to encourage western settlement in the nineteenth century, were exceptions. William Ferrel, the most mathematically astute geophysicist in America and a theorist of the general circulation of the Earth's oceans and atmosphere, argued that increased rainfall could not be directly attributed to the presence of forests or cultivated fields. Basing his arguments on continental and planetary scale factors rather than local conditions, Ferrel reasoned that abundant rainfall depended on evaporation at a distance, and that any increased water vapor due to local influences "would be carried so far and spread over so great a territory, that the increased rainfall at any given place would perhaps be entirely insensitive to observation."[19]

COSMIC PHYSICS AND THE BRIEF RETURN
OF HUMAN INFLUENCE

The piecemeal construction of "geophysics" as a discipline in the nineteenth century—a tradition that privileged cosmic physics and generally excluded human agency—was a result of the combined efforts of many scholars.[20] The work of Joseph Fourier, John Tyndall, and Svante Arrhenius, representing respectively the early, middle, and late decades of the nineteenth century, illustrate this tradition in global climate studies and form elements of an intellectual ancestry of the greenhouse effect. Arrhenius ultimately reintroduced humans as possible global geophysical agents of climate change.

Jean Baptiste Joseph Fourier (1768–1830) is remembered today as a mathematical physicist, but was known to his contemporaries as variously, a secret policeman, a political prisoner (twice), governor of Egypt, prefect of Isère and Rhône, friend of Napoleon, baron, outcast, and perpetual member and secretary of the French Academy of Sciences, whose fortunes rose and fell with the swirling political tides.[21] Fourier considered himself the Newton of heat: "The principle of heat penetrates, like gravity, all objects and all of space, and it is subject to simple and constant laws."[22] For Fourier, solar heating and the temperature of outer space were the most significant factors controlling the Earth's heat budget,[23] although he did write suggestively about what others later called the greenhouse effect. For example: "the temperature [of the Earth] can be augmented by the interposition of the atmosphere, because heat in the state of light finds less resistance in penetrating the air, than in repassing into the air when converted into non-luminous heat."[24] Yet the author of the *Analytical Theory of Heat* admitted that it is "difficult to know how far the atmosphere influences the mean temperature of the globe; and in this examination we are no longer guided by a regular mathematical theory."[25]

In the middle decades of the nineteenth century, John Tyndall (1820–1893), Irish-born scientist and consummate experimentalist, worked at the Royal Institution on the radiative properties of various gases, demonstrating that "perfectly colorless and invisible gases and vapours" were able to absorb and emit radiant heat. He identified the importance of atmospheric trace constituents as efficient absorbers of long wave radiation and as important factors in climatic control. Specifically, he established beyond a doubt that the radiative properties of water vapor and carbon dioxide were of importance in explaining meteorological phenomena such as the formation of dew, the energy of the solar spectrum, and possibly the variation of climates over geological time.

As an accomplished lecturer and writer, Tyndall employed numerous metaphors to describe his experiments with radiant heat. Here are two reminiscent of the greenhouse effect: "The solar heat possesses . . . the power of crossing an atmosphere; but, when the heat is absorbed by the planet, it is so changed in quality that the rays emanating from the planet cannot get with the same freedom back into space. Thus the atmosphere admits of the entrance of the solar heat, but checks its exit; and the result is a tendency to accumulate heat at the surface of the planet."[26] In an alternative greenhouse metaphor, Tyndall wrote: "The aqueous vapour constitutes a local dam, by which the temperature at the earth's surface is deepened; the dam, however, finally overflows, and we give to space all that we receive from the sun."[27]

The carbon dioxide content of human lungs was also a focus of Tyndall's investigations. In 1864, W.F. Barrett conducted experiments on the "carbonic acid contained in our breath." He did so at Tyndall's request using his apparatus at the Royal Institution for investigating the radiative properties of gases. Barrett filled vulcanized India rubber bags with samples of his own breath, "about a half an hour after rising," "about 10 minutes after breakfast," and "after a brisk walk." The bags were analyzed by Dr. Edward Frankland, one of the leading chemists of London, and the results—including a fourth bag from Frankland's lungs, collected after "considerable exertion"—were compared with Tyndall's longwave radiation absorption method. The percentage of carbonic acid found by the two techniques varied by only 0.1 to 0.3 percent, confirming for Barrett, "the correctness of the principle and the general accuracy of the observations." When it was discovered that the amount of absorption varied on different days, in different weather conditions, and when air from different source regions was tested, Barrett announced excitedly, "it is here evident that the variations in the amount of carbonic acid in the atmosphere can be detected by this mode of experiment." Thus was suggested an early program to measure spatial and temporal variations in the concentration of carbon dioxide in the free atmosphere as well as in human lungs.[28]

On a more cosmic level, Tyndall thought that changes in the amount of any of the radiatively active constituents of the atmosphere—water vapor, carbon dioxide, ozone, or hydrocarbons—could have produced *"all the mutations of climate which the researches of geologists reveal* . . . they constitute true causes, the *extent* alone of the operation remaining doubtful." Tyndall gave credit to his predecessors, including Fourier, for the intuition that "the rays from the sun and fixed stars could reach the earth through the atmosphere more easily than the rays emanating from the earth could get back into space. The experimental verification of this phenomenon, however, belonged to Tyndall.[29] His laboratory experiments with microphysical entities on Albemarle Street had informed a new view of the cosmos.

In 1896, Svante Arrhenius (1859–1927), following Tyndall's suggestion, demonstrated that variations of atmospheric CO_2 concentration could have a very great effect on the overall heat budget and surface temperature of the planet and might trigger feedback phenomena that could account for glacial advances and retreats.[30] For Arrhenius, who adopted a carbon budget developed by his colleague Arvid Högbom, volcanoes were the chief source of carbon dioxide in the Earth's atmosphere.

The theory of human influence on climate returned briefly in 1899 when Nils Ekholm pointed out that at present rates, the burning of pit coal could double the concentration of atmospheric CO_2 and could "undoubtedly cause a very obvious rise of the mean temperature of the Earth." In his book *Worlds in the Making*, Arrhenius popularized Ekholm's observation, noting that "the slight percentage of carbonic acid in the atmosphere may by the advances of industry be changed to a noticeable degree in the course of a few centuries." Arrhenius considered it likely that in future geological ages the Earth would be "visited by a new ice period that will drive us from our temperate countries into the hotter climates of Africa." On the time scale of hundreds to thousands of years, however, Arrhenius speculated on a "virtuous circle" in which the burning of fossil fuels could help prevent a rapid return to the conditions of an ice age, and could perhaps inaugurate a new carboniferous age of enormous plant growth. Ekholm concurred, adding his own speculations about the possibility of deliberate climate modification by burning coal exposed in shallow seams.[31]

ECLIPSE AND RE-EMERGENCE
OF HUMAN INFLUENCE

In the first half of the twentieth century, most scientists did not believe that increased levels of atmospheric carbon dioxide would result in global warming. It was thought that at current atmospheric concentrations, the gas already absorbed all the available long-wave radiation; thus any increases would not change the radiative heat balance of the planet but might augment plant growth. Other physical mechanisms of climatic change, although highly speculative, were given more credence, especially changes in solar luminosity, atmospheric transparency, and the Earth's orbital elements.

In his articles on climate change, beginning in 1938, G.S. Callendar (1898–1964), a leading British steam engineer, acknowledged the "chequered history" of the CO_2 theory: "[I]t was abandoned for many years when the prepondering influence of water vapour radiation in the lower atmosphere was first discovered, but was revived again a few years ago when more accurate measurements of the water vapour spectrum became available." Noting that humans had long been able to intervene in and accelerate natural processes, Callendar pointed out that humanity was now intervening heavily in the slow-moving carbon cycle by "throwing some 9,000 tons of carbon dioxide into the air each minute." He revived and reformulated the CO_2 theory by arguing that rising global temperatures and increased fuel burning were closely linked.[32]

Callendar was employed as an engineer in British defense research and worked on the problem of climate change in his spare time. He recorded his own highly accurate series of daily weather observations from 1942 to 1964 and also collected and compiled weather records and trends from stations around the world, filling dozens of research notebooks in the process. The pattern he identified clearly indicated a pronounced global warming trend in the early decades of the twentieth century.

Callendar investigated the carbon cycle, including natural and anthropogenic sources and sinks, and the role of glaciers in the Earth's heat budget. From his review of earlier measurements, Callendar established the now standard number of 290 parts per million (ppm) as the nineteenth-century background concentration of carbon dioxide in the atmosphere. He documented an increase of ten percent in this figure between 1900 and 1935, which closely matched the amount of fuel burned in this era.

Using newly-available and detailed spectroscopic data, Callendar formulated a theory of infrared absorption and emission by trace gases and argued that the rising carbon dioxide content of the atmosphere and the rising temperature were due to anthropogenic activities. In an era before computer climate modeling, Callendar compiled all the newly available information on the CO_2 spectrum into a coherent picture of interest and relevance to meteorologists, thus establishing the carbon dioxide theory of climate change in its recognizably modern form and reviving it from its earlier, physically unrealistic and moribund status. The notable series of articles on climate change he published between 1938 and 1961 were the result of his extensive calculations, laboratory measurements, literature searches, and professional correspondence.

By the 1950s, as temperatures around the Northern Hemisphere reached early-twentieth-century peaks, global warming first found its way onto the public agenda. Concerns were expressed in both the scientific and popular press about rising sea levels, loss of habitat, and shifting agricultural zones. Amid the myriad mechanisms that could possibly account for climatic changes, G. S. Callendar was the first of many to document connections between rising surface temperatures, increasing anthropogenic CO_2 emissions, infrared radiation, and global climate warming. His writings revived the theme of human agency, which had been dormant since the age of Jefferson, pointing out that humanity had sped up natural processes and had become an agent of global change by interfering with the carbon cycle. For Callendar, the rise in global temperature in the early decades of the twentieth century was inextricably linked to the concurrent increase in industrial emissions of carbon dioxide.

Today the "Callendar Effect"[33] refers to the theory that global climate change can be attributed to an enhanced greenhouse effect due to elevated levels of carbon dioxide in the atmosphere from anthropogenic sources, primarily from the combustion of fossil fuels. Callendar wrote in 1939, "As man is now changing the composition of the atmosphere at a rate which must be very exceptional on the geological time scale, it is natural to seek for the probable effects of such a change. From the best laboratory observations it appears that the principal result of increasing atmospheric carbon dioxide . . . would be a gradual increase in the mean temperature of the colder regions of the earth."[34] The now famous mantra that humans are "carrying out a large scale geophysical experiment" was also foreshadowed by Callendar, when he wrote in 1938, "The course of world temperatures during the next twenty years should afford valuable evidence as to the accuracy of the calculated effect of atmospheric carbon dioxide."[35] Although between 1940 and 1970 global temperatures showed little trend, even a slight cooling, today most scientists would agree that these early statements have been verified. One author recently referred to Callendar's insights as being "pretty much spot-on" and his work on the enhanced greenhouse effect and human agency as "roughly half a century ahead of his time."[36]

TENTATIVE PLANS TO INTERVENE

In 1979, based in part on early results from computer climate models, the Ad Hoc Study Group on Carbon Dioxide and Climate of the National Academy of Sciences reported, "If carbon dioxide continues to increase, the study group finds no reason to doubt that climate changes will result and no reason to believe that these changes will be negligible." They advised against a "wait-and-see policy," since the oceans would likely mask climate warming until it was too late to respond.[37] After the passage of almost fifteen years, and a large number of studies and reports later, the IPCC Second Assessment Report (1995) stated judiciously, "the balance of evidence suggests a discernible human influence on global climate"; six years later the US National Academy of Sciences echoed this statement: "Greenhouse gases are accumulating in Earth's atmosphere as a result of human activities, causing surface air temperatures and subsurface ocean temperatures to rise," with the caveat that "we cannot rule out that some significant part of these changes are also a reflection of natural variability."[38]

Assuming that adverse climate change impacts may be greater than anticipated and carbon reduction measures may be less effective than anticipated, increasing consideration has been given in recent decades to possible, but highly controversial macro-engineering options for the management and mitigation of climate change. As an early example of this, *Restoring the Quality of Our Environment* (1965), issued by the President's Science Advisory Committee, provided estimates of the future increase of anthropogenic CO_2 due to fossil fuel burning and its likely negative impact on climate, concluding that geoengineering options, or as they put it, "the possibilities of deliberately bringing about countervailing climatic changes . . . need to be thoroughly explored." As an illustration, the committee pointed out that the Earth's albedo could be increased by one percent by dispersing buoyant reflective particles on the sea surface at an annual cost, not considered excessive, of about $500 million. Reducing fossil fuel use was not mentioned as an option.[39]

During the hot summer of 1988, the Government of Canada hosted a major international scientific conference on "The Changing Atmosphere: Implications for Global Security" in collaboration with the United Nations and the World Meteorological Programme. By conference end, scientists from all over the world agreed on a consensus statement and a target for emission reductions. The statement opened, "Humanity is conducting an unintended, uncontrolled, globally pervasive experiment, whose ultimate consequences could be second only to global nuclear war." The target: global reductions of carbon dioxide emissions twenty percent below 1988 levels to be achieved by 2005. That was 1988. In 2005 we are nowhere near reaching this goal, while popular cries of "Stop Global Warming" and "Control Climate Change" are becoming more and more widespread.

Policy Implications of Greenhouse Warming (1992), a report from the National Academy of Science, is predicated on the idea that "Global warming continues to gain importance on the international agenda and calls for action are heightening. Yet, there is still controversy over what must be done

FIGURE 3   Shooting dust into the stratosphere to offset global warming. Cartoonist's interpretation of 1992 National Academy proposal, from *Geographic Magazine* (May 1992), 23. Nobel Laureate Paul J. Crutzen revived this proposal in 2006 (note 45).

and what is needed to proceed." One of their policy recommendations was that the U.S. should conduct research in schemes to cool the Earth if global warming gets out of hand. Proposals included orbiting a fleet of space mirrors or shooting dust or spraying sulfur dioxide into the stratosphere to reflect solar radiation back to space, turning the oceans into soupy green algae blooms to sequester excess carbon, or setting up gigantic "soot generators" to shade the Earth.[40]

CLIMATE ENGINEERING IN
THE TWENTY-FIRST CENTURY

Geoengineering, or purposeful, large-scale manipulation of the environment, has been linked to climatic or hydrological changes thought to be of widespread benefit, at least to their proposers. Today's geoengineering schemes are typically ocean-based (diverting currents, iron fertilization, ocean carbon sequestration), land-based (biological or geological carbon sequestration, alternative energy generation), or radiation-based (space mirrors, enhancing cloud reflectivity, eliminating trace gases).

In October 2003, the U.S. Pentagon released a controversial report titled "An Abrupt Climate Change Scenario and Its Implications for United States National Security," explaining how global warming could lead to rapid and catastrophic global cooling and recommending that the government "explore geo-engineering options that control the climate."

> Today, it is easier to warm than to cool the climate, so it might be possible to add various gases, such as hydro-fluorocarbons, to the atmosphere to offset the affects (sic.) of cooling. Such actions, of course, would be studied carefully, as they have the potential to exacerbate conflicts among nations.[41]

In January 2004, a symposium on "Macro-engineering Options for Climate Change Management and Mitigation" was held in Cambridge, England under the joint sponsorship of the Tyndall Centre for Climate Change Research and the business-oriented Cambridge-MIT Institute. The organizers cited as their rationale the urgent need to reduce greenhouse emissions by fifty percent globally and up to ninety percent in the United States and Europe in order to avoid excessive climate change, and the *unlikelihood* of such reductions being accomplished by conventional means such as renewable energy sources and energy efficiency. The participants set out to "identify, debate, and evaluate" possible, but highly controversial macro-engineering options. This was no mere academic exercise, but a fully vested rehearsal, ranking, and evaluation by the research community and their government sponsors of the panoply of geoengineering options just prior to their implementation.

Although couched in the language of uncertainty and swathed in caveats, the Tyndall Centre conference coincided with the initiation of actual pilot projects and served to move the speculative geoengineering agenda closer to the mainstream. In the language of the organizers, "At the very least, [these] may be considered as emergency policy options in the event of more adverse climate change impacts than expected, or less effective carbon reduction measures than anticipated." Note that "adverse climate change impacts" were not specified and less-than-effective carbon reduction measures are just about certain to occur. Surely macro-scale or planetary-scale climate engineering—the study, preparation, and execution of the largest possible engineering works—also requires macrosocial planning, implying fundamental changes to the world's economic and political systems, social and cultural institutions, and even ethnic and demographic groupings. Yet the macro-engineers had gathered—without the collective perspectives of historians, ethicists, or social scientists![2]

Current assessments of weather modification include the ability to clear cold fogs, suppress frost, and augment orographic precipitation by up to ten percent, but none of the more extravagant claims, including drought relief, storm modification, weather warfare, or climate control have materialized. That has not deterred NASA's Institute for Advanced Concepts from funding Ross Hoffman's research on microwaving hurricanes to redirect them—assuming one knew where the storm was originally headed and that there would be no liabilities along its new path! Nor has it deterred Senator Kay Bailey Hutchison of Texas from introducing the Weather Modification Research and Technology Transfer Authorization Act of 2005.

Just how much climate change would be needed to trigger a geoengineering option? And who would make these decisions for the world? One candidate, now deceased, was Edward Teller, who announced in the press that "The Planet Needs a Sunscreen." In subsequent articles, including his last posthumous one for the Tyndall Centre symposium, Teller and his protégés argued that "active technical management of radiative forcing" (or albedo engineering) was needed to prevent both global warming and ice ages, and that technical fixes were vastly cheaper and universally preferable to bureaucratic controls or lifestyle changes. Moreover, the UN Framework Convention on Climate Change mandated them![43] The authors' proposal to control global warming by injecting reflective particles into the stratosphere would have the effect of turning the blue sky white while reducing direct beam solar radiation by about twenty percent.[44] Perhaps it would also generate more colorful sunsets!

Lest you think this is all fantasy, some of the leading scientists in the world are today proposing, or at least seriously considering, the implementation of geoengineering schemes. In November 2005, Russian scientist Yuri Israel, the head of the Global Climate and Ecology Institute, wrote to President Putin that global warming requires immediate action and suggested burning thousands of tons of sulfur in the stratosphere to lower the earth's temperature "a degree or two"—note that this is more than the total climate warming since pre-industrial times. According to Israel's optimistic assessment, "we really will be able to control the climate" with this "ecologically safe" remedy. Also in July 2006, Nobel Laureate and atmospheric chemist Paul J. Crutzen published an editorial containing similar thoughts on stratospheric sulfur injection.[45] If this does not raise your apprehensions, I don't know what will.

CONCLUSION

To date, the discourse on geoengineering has largely been speculative and pragmatic, based on risk assessment and cost-benefit analysis, with serious historical analysis and ethical arguments about geoengineering almost non-existent. A large-scale environmental technological fix cast as an response to undesired climate change could be seen as an act imposed on the multitude by the will of the few, for the primary benefit of those already in power. Some would undoubtedly interpret it as a hostile or aggressive act. Moreover, the ability to use technology to counteract the effects of greenhouse gas emissions contains the moral trap of eliminating incentives for conservation and clean energy generation. Rather than engaging in speculative large-scale climate engineering, it is better to reduce the effects of greenhouse gas emissions—by reducing greenhouse gas emissions.[46]

In climate modification or in any other field, broader, historical perspectives are needed, to inform ethical analysis and sound public policy decisions. The goal is the articulation of a perspective fully informed by history and the initiation of a dialogue that uncovers otherwise hidden values, ethical implications, social tensions, and public apprehensions about the large-scale control of nature. A Union session of four papers at the American Geophysical Union in December 2005 examined geoengineering history, ethics, and policy, but, I would suggest, more work on this subject is essential.[47]

Hansen may be right: we may be approaching the physical tipping point of climate, or, as James Lovelock argues in his new book, *The Revenge of Gaia*, we may already have passed it, with catastrophic consequences for humanity.[48] More likely, Hansen, Lovelock, Al Gore, and many others are trying to add their weight to the sociological tipping point that convinces all of humanity to take concerted action against inadvertently harming the climate system.

> As the effects of global warming become more and more apparent, will we react by finally fashioning a global response? Or will we retreat into ever narrower and more destructive forms of self-interest? It may seem impossible to imagine that a technologically advanced society could choose, in essence, to destroy itself, but that is what we are now in the process of doing.[49]

One way of taking action, although exceedingly difficult, is to attempt to return to a lower level of carbon dioxide in the atmosphere through treaties, conservation, and clean energy generation. Even the reasonable and seemingly manageable "stabilization wedges" of Stephen Pascala and Robert Socolow—fifteen strategies to enhance energy efficiency, reduce carbon emissions, and sequester CO_2—may not be politically feasible since they call for actions that go well beyond business as usual.[50]

Unfortunately, extreme apocalyptic perspectives on climate change may lead to another kind of tipping point, in which elites decide it is in the best interest of humanity to exercise the geoengineering option. This would pose all sorts of ethical problems and the moral dilemma that if at all successful, climate modification would serve as a *disincentive* to clean energy practices. In 1969, Joseph Fletcher wrote, "It is convenient to think of progress toward climate control in four stages—observation, understanding, prediction, and control. We must observe *how* nature behaves before we can understand *why*, we must *understand* before we can *predict*, and we must be able to *predict* the outcome before we undertake measures for *control*."[51] Fletcher's advice is sage but insufficient. Clearly, in making policy decisions involving geoengineering, purely technical proposals and back-of-the-envelope calculations are not good enough. The climate community must include in all of its panels an interdisciplinary mix of scientists, historians, ethicists, and policy makers; it would be well served to avoid Archimedean-scale interventions and follow the Hippocratic prescription for a planetary fever: "Declare the past, diagnose the present, foretell the future; practice these acts. As to diseases, make a habit of two things—to help, or at least to do no harm."[52] The alternative is invasive planetary surgery.

NOTES

1. James Hansen, "The Tipping Point?" *New York Review of Books* 53, 1 (12 Jan. 2006), http://www.nybooks.com/articles/18618 (3 Jan. 2006). The term was originally used by sociologists to describe the threshold moment when a change in collective behavior begins to occur; for example, *Scientific American* (Oct. 1957): 34. Recently, the term "tipping point" has been applied to positive feedback processes in which, beyond a certain point, the rate of climate change increases dramatically or equilibrium breaks down, resulting in "ecological changes from which the world cannot recover." Ian Sample, "When will global warming reach a point of no return?" *The Guardian* (27 Jan. 2005).
2. *Time* (3 April 2006); Mark Hertsgaard, "While Washington Slept," *Vanity Fair* (May 2006): 200–07. Climate apocalypse is also a theme emphasized in Al Gore's 2006 film, *An Inconvenient Truth* ("By far, the most terrifying film you will ever see!").
3. On intervention, see James Rodger Fleming, "Fixing the Weather and Climate: Military and Civilian Schemes for Cloud Seeding and Climate Engineering," *The Technological Fix: How people use technology to create and solve problems*, Lisa Rosner, ed. (New York: Routledge, 2004), 175–200; and James Rodger Fleming, "The Pathological History of Weather and Climate Modification: Three cycles of promise and hype," *Historical Studies in the Physical Sciences* (2006), in press.
4. An interesting article on ethics, but with no mention of geoengineering, is Stephen Gardiner, "Ethics and Global Climate Change," *Ethics* 114 (2004): 555–600. Much of the review literature on human agency focuses on "inadvertent" influences since the second half of the twentieth century: Carroll L. Wilson and William H. Matthews, eds., *Inadvertent Climate Modification. Report of Conference, Study of Man's Impact on Climate (SMIC), Stockholm* (Cambridge, MA: MIT Press, 1971); Earl W. Barrett and Helmut E. Landsberg, "Inadvertent Weather and Climate Modification," *CRC Critical Reviews in Environmental Control* 6 (1975): 15–90; William W. Kellogg, "Mankind's Impact on Climate: The Evolution of an Awareness," *Climatic Change* 10 (1987):

113–36; David M. Hart and David G. Victor, "Scientific Elites and the Making of US Policy for Climate Change Research, 1957–74," *Soc. Stud. Sci.* 23 (1993): 643–80; and Spencer Weart, "The Discovery of Global Warming," http://www .aip.org/history/climate/index.html (17 Jan. 2006).

5. This was one of the basic questions addressed in James Rodger Fleming, *Historical Perspectives on Climate Change* (New York: Oxford University Press, 1998), 7–8. Here I provide new perspectives and bring the story up to date.

6. Clarence Glacken, *Traces on the Rhodian Shore: Nature and culture in Western thought from ancient times to the end of the eighteenth century* (Berkeley: University of California Press, 1967), iii–v. The other two perennial questions involved teleology and determinism.

7. James Rodger Fleming, "A History of Weather and Climate Control," AAAS Roger Revelle Fellowship Proposal, 2006.

8. Theophrastus, cited in Glacken, *Traces*, 130.

9. Ibid.

10. Montesquieu, *The Spirit of Laws: A compendium of the first English edition*, D. W. Carrithers, ed. (Berkeley: University of California Press, 1977), 50.

11. David Hume, "Of the Populousness of Ancient Nations" (ca. 1750), in Hume, *Essays: Moral, Political, and Literary*, T. H. Green and T. H. Grose, eds. (London, 1875), 1: 434.

12. This is well documented. For details see James Rodger Fleming, *Meteorology in America, 1800–1870* (Baltimore: Johns Hopkins University Press, 1990), 2–3; and Karen Ordahl Kupperman, "The Puzzle of the American Climate in the Early Colonial Period," *Amer. Hist. Rev.* 87 (1982): 1270.

13. Hugh Williamson, *Observations on the Climate in Different Parts of America, compared with the climate in corresponding parts of the other continent . . .* (New York, 1811).

14. Thomas Jefferson to Lewis E. Beck, July 16, 1824, in Albert Ellery Bergh (ed.), *The Writings of Thomas Jefferson*, XV (Washington, D.C.: Thomas Jefferson Memorial Association of the United States, 1907), 71–72.

15. Fleming, *Meteorology in America*, 9–20 and passim.

16. Samuel Forry, "Researches in Elucidation of the Distribution of Heat over the Globe, and Especially of the Climatic Features Peculiar to the Region of the United States," *Amer. J. Sci. Arts* 47 (1844): 236; and Forry, *The Climate of the United States and Its Endemic Influences* (New York, 1842).

17. Fleming, *Meteorology in America*, 15, 68–69, 160–61.

18. Henry Nash Smith, "Rain Follows the Plow: The Notion of Increased Rainfall for the Great Plains, 1844–1880," *Huntington Library Quarterly* 10 (1947): 169–93.

19. William Ferrel, "Note on the Influence of Forests Upon Rainfall," *Amer. Meteorol. J.* 5 (1888–89): 433–35.

20. Gregory A. Good, "The Assembly of Geophysics: Scientific Disciplines as Frameworks of Consensus," *Perspectives on Geophysics*. Special theme issue of *Stud. Hist. Phil. Mod. Phys.*, vol. 31, no. 3 (2000): 259–92.

21. I. Grattan-Guinness, with J. Ravitz, *Joseph Fourier, 1768–1830: A survey of his life and work, based on a critical edition of his monograph on the propagation of heat presented to the Institute of France in 1807* (Cambridge: MIT Press, 1972); and John Herivel, *Joseph Fourier: The man and the physicist* (Oxford: Clarendon Press, 1975).

22. Joseph Fourier, "Extrait d'une mémoire sur l'état actual de la théorie physique et mathematique de chaleur," n.d., Théorie de la chaleur 7. Ouvrages sur la chaleur, MS, Collection des papiers du mathématicién Fourier 29 (MSS français 22529), 79, Bibliothèque nationale, Paris.

23. Joseph Fourier, "Remarques générales sur les températures du globe terrestre et des espaces planétaires," *Ann. Chim. Phys.* 2nd ser., 27 (1824): 136–67. English translation by Ebeneser Burgess in *Amer. J. Sci. Arts* 32 (1837): 1–20.

24. Ibid., 155; Burgess translation, 13.

25. Ibid., 151–53; Burgess translation, 10–11.

26. John Tyndall, "On the Transmission of Heat of Different Qualities Through Gases of Different Kinds," *Proc. Roy. Inst. Gt. Br.* 3 (1858–62): 158.

27. John Tyndall, "On Radiation through the Earth's Atmosphere," Friday, 23 Jan. 1863, *Proc. Roy. Inst. Gt. Br.* 4 (1851–66): 4–8; quote from 8; also in *Phil. Mag.* ser. 4, 25 (1863): 200–206.

28. W.F. Barrett, "On a Physical Analysis of the Human Breath," *Phil. Mag.* 28 (1864): 108–21. Spatial and temporal observation of carbon dioxide concentrations continued in the twentieth century.

29. Fleming, *Historical Perspectives on Climate Change*, 68–71.

30. Svante Arrhenius, "On the Influence of Carbonic Acid in the Air upon the Temperature of the Ground," *Phil. Mag.*, ser. 5 (1896): 237–76; Elisabeth Crawford, *Arrhenius: From Ionic Theory to the Greenhouse Effect* (Canton, MA: Science History Publications, 1996).

31. Nils Ekholm, "On the Variations of the Climate of the Geological and Historical Past and Their Causes," *Quart. J. Roy. Meteorol. Soc.* 27 (1901): 61. This article appeared in Swedish in 1899 and was translated two years later. Svante Arrhenius, *Worlds in the Making: The evolution of the universe*, transl. H. Borns (New York: Harper, 1908), 54–63; Fleming, *Historical Perspectives on Climate Change*, 82, 111; David Keith, "Geoengineering & Climate: An Overview," unpublished paper presented at the Tyndall Centre conference on Macro-engineering Options for Climate Change Management & Mitigation, Cambridge, UK, Jan. 2004.

32. G.S. Callendar, "The Artificial Production of Carbon Dioxide and Its Influence on Temperature," *Quart. J. Roy. Meteorol. Soc.* 64 (1938): 223–240; G.S. Callendar, "Can Carbon Dioxide Influence Climate?" *Weather* 4 (1949): 310; G.S.

Callendar, "The Composition of the Atmosphere through the Ages," *Meteorol. Mag.* 74 (1939): 38.

33. Callendar's biography has just been published and his papers are available in a digital edition. James Rodger Fleming, *The Callendar Effect: The life and work of Guy Stewart Callendar (1898–1964), the British scientist who established the carbon dioxide theory of climate change* (Boston: American Meteorological Society, 2007); issued with an optional DVD of the *The Callendar Papers, Digital edition*, edited by James Rodger Fleming and Jason Thomas Fleming (Boston: American Meteorological Society, 2007).

34. Callendar, "Composition of the Atmosphere," 38

35. Callendar, "Artificial Production of Carbon Dioxide," 236.

36. Mark Bowen, *Thin Ice: Unlocking the secrets of the world's highest mountains* (New York: Henry Holt, 2005), 96.

37. *Carbon Dioxide and Climate: A scientific assessment* (National Academy Press, Washington, DC: 1979). For a guide to U.S. climate change policy from 1978 to 1992 see Wayne A. Morrissey, "Global Climate Change: A Survey of Scientific Research and Policy Reports," *Congressional Research Service Report for Congress*, RL30522 (Washington, DC: Library of Congress, 2000).

38. Intergovernmental Panel on Climate Change, Second Assessment Report, 1995, http://www.ipcc.ch/pub/sarsum1.htm (19 Jan. 2006); U.S. National Academy of Science, *Climate Change Science: An analysis of some key questions* (Washington, DC, 2001).

39. President's Science Advisory Council, *Restoring the Quality of Our Environment*, Report of the environmental pollution panel (Washington, DC, 1965).

40. National Academy of Sciences, *Policy Implications of Greenhouse Warming: Mitigation, adaptation, and the science base* (Washington, DC, 1992).

41. Peter Schwartz and Doug Randall. "An Abrupt Climate Change Scenario and Its Implications for United States National Security." 2003. http://www.environmentaldefense.org/documents/3566_AbruptClimateChange.pdf (2 Jan. 2006).

42. David Keith, a lone social scientist at the Tyndall Centre conference, was of the opinion that "it is nearly inevitable that we will come to actively manage, even engineer, planetary scale processes." Citation from typescript remarks.

43. Edward Teller, "The Planet Needs a Sunscreen," *Wall Street Journal* (17 Oct. 1997), 1; Edward Teller, et al., "Active climate stabilization: Presently-feasible albedo-control approaches to prevention of both types of climate change" in Tyndall Centre conference, http://www.tyndall.ac.uk/events/past_events/cmi.shtml; Framework Convention on Climate Change, Article 3.3.

44. Personal communication between author and Michael MacCracken, former director of the U.S. Global Change Research Program (12 Feb. 2005).

45. "Russian Scientist Suggests Burning Sulfur in Stratosphere to Fight Global Warming," *MosNews* (30 Nov. 2005), http://www.mosnews.com/news/2005/

11/30/brimstoneskies.shtml. Paul J. Crutzen, "Albedo Enhancement by Strato-spheric Sulfur Injections: A contribution to resolve a policy dilemma?" *Climatic Change* (25 July 2006).

46. See D. Whitney King, "Can Adding Iron to the Oceans Reduce Global Warm-ing? An example of geoengineering," James Rodger Fleming and Henry A. Gemery, eds., *Technology and the Environment: Multidisciplinary perspectives* (Akron: Akron University Press, 1994), 112–35.

47. James R. Fleming, convener, "Geoengineering: Historical, ethical, and policy perspectives," American Geophysical Union Session U54A, San Francisco (9 Dec. 2005) and proposer, "Sustaining the Global Climate: Science, ethics, and pub-lic policy," AAAS Annual Meeting, San Francisco, 2007.

48. James Lovelock, *The Revenge of Gaia* (New York: Penguin, 2006); see Love-lock, "The Earth is about to catch a morbid fever that may last as long as 100,000 years, *The Independent* (16 Jan. 2006).
http://comment.independent.co.uk/commentators/article338830.ece

49. Elizabeth Kolbert, "The Climate of Man," I: "Disappearing islands, thawing permafrost, melting polar ice? How the Earth is changing," II: "The curse of Akkad," III: "What can be done?" *New Yorker* (25 April, 2 May, 9 May, 2005).

50. Stephen Pascala and Robert Socolow, "Stabilization Wedges: Solving the climate problem for the next 50 years with current technologies," *Science* 305 (2004): 968–72.

51. J.O. Fletcher, "Managing Climatic Resources," RAND Report No. P-4000-1, 1969.

52. Hippocrates, *Epidemics* I: 11.

Notes on Contributors

Katharine Anderson, Associate Professor, Science & Technology Studies, York University, Toronto, earned her Ph.D. in History from Northwestern University. She published *Predicting the Weather: Victorians and the Science of Meteorology* with the University of Chicago Press in 2005 and is currently studying ocean sciences in the early twentieth century. In 2005–06 she was on leave and a fellow at Massey College, Toronto.

Deborah R. Coen is Assistant Professor of History at Barnard College, Columbia University. She received her Ph.D. in History of Science from Harvard University and was a Junior Fellow of the Harvard Society of Fellows. She is the author of *Vienna in the Age of Uncertainty: Science, Liberalism, and Private Life*, to be published by University of Chicago Press in 2007.

Gregory T. Cushman, is Assistant Professor of International Environmental History at the University of Kansas and received his Ph.D. in Latin American history from the University of Texas at Austin in 2003. He is currently completing a book titled *The Lords of Guano: Global Ecology and Peru's Marine Environment* and has embarked on an investigation of the history of understanding of the tropical ocean and atmosphere focused on the El Niño phenomenon.

James Rodger (Jim) Fleming, Professor of Science, Technology and Society at Colby College, earned his Ph.D. in History from Princeton University. His books include *Meteorology in America, 1800–1870* (Johns Hopkins, 1990), *Historical Perspectives on Climate Change* (Oxford, 1998), and *The Cal-*

lendar Effect (American Meteorological Society, 2007). Current projects include an illustrated history of "killer apps, cutting edges, and tipping points" in meteorology and a comprehensive history of weather and climate control. He currently holds the Roger Revelle Fellowship in Global Stewardship from the American Association for the Advancement of Science. He is a Public Policy Scholar with the Woodrow Wilson International Center for Scholars and a visiting scholar with the U.S. National Academy of Sciences and the Smithsonian National Air and Space Museum.

Gregory A. Good, Associate Professor in History of Science, West Virginia University, earned his Ph.D. in History of Science at the Institute for the History and Philosophy of Science and Technology at the University of Toronto, Canada. He served six years as editor of *Earth Sciences History* and has published two edited books: *The Earth, the Heavens, and the Carnegie Institution of Washington* (American Geophysical Union, 1994) and *Sciences of the Earth: An Encyclopedia of Events, People, and Phenomena* (Garland, 1998). He is currently writing "Magnetic World," a history of geomagnetic research in the 19th and 20th centuries, supported by NSF grant 0432202. In 2005–06 he was a Visiting Scholar at the University of Cambridge, UK, Department of the History and Philosophy of Science and Clare Hall.

Vladimir Jankovic is Wellcome Research Lecturer at the Centre for the History of Science, Technology and Medicine at the University of Manchester. He received his Ph.D. from the University of Notre Dame and is the author of *Reading the Skies: A cultural history of English weather, 1650–1820* (Chicago, 2000). He is completing a Wellcome research project on the medical history of indoors and its impact on modern environmental sensibilities. He was also a consultant and presenter for a Discovery Channel series *The Storms of War*.

Richard Staley, Assistant Professor, History of Science, University of Wisconsin–Madison; B.A., University of Melbourne; Ph.D., University of Cambridge, has broad-ranging interests in the history of physics in Europe and America during the 19th and 20th centuries, with a particular research focus on the physics community circa 1900, the development of special relativity, and the interrelations between instruments and experiment.

Roger Turner is a graduate student at the University of Pennsylvania, where he explores the history of science, technology and the environment. His dissertation will examine the intertwined development of meteorology, aviation, and electronics during the twentieth century, investigating how novel machines (like aircraft, radar, computers, and satellites) directed the attention of scientists to new phenomena, while meteorological knowledge simultaneously shaped the design and operation of those machines. As a reader, he's fascinated by the tragic romance of dirigibles.

Index of Names

AAF Weather Service, 157
Abbe, Cleveland, 232
Abbott, Charles Greely, 147
Abercromby, Ralph, 72
Aberdeen, 20
"Abrupt Climate Change Scenario and
    Its Implications for United States
    National Security, An," 240
Adair, James Makittrick, 13, 14, 18
Ad Hoc Study Group on Carbon
    Dioxide and Climate, 238
Admiralty, 48, 55
Admiralty Manual (Herschel), 46, 55
Aéropostale, 190
Africa, 58, 229, 235
Age of Aviation, 178
Air Commerce Act of 1926, 153
Air Services Branch (Canada), 194
Airy, George Biddell, 51, 57, 72
Aitken, John, 95, 101, 102, 103, 107,
    108
Akron (Navy dirigible), 145, 154
Alameda Park, Montevideo, 179
Albertus Magnus, 228
"Alpine Weather Service," 123
Alps, xiii, 6, 116, 120, 124, 125, 127,
    128, 130
America, 12, 35, 54, 75, 148, 230
American Geophysical Union, 223, 243

American Meteorological Society, 117,
    146, 163, 193, 199
American Scandinavian
    Foundation/fellowship, 144, 149,
    150
Americas, the, 2, 58, 177, 178, 184,
    186, 187, 230
AMS *Bulletin*, 199
Analytical Theory of Heat (Fourier),
    233
Anderson, Katharine, xiv, xv, 69
Andover, 13
Anthropogenic climate change, 226
Antigua, 13
Arago, Françoise, 44, 56
Arbuthnot, John, 228
Archimedes, 225
Arctic, the, 191
Arctic Ocean, 175
Argentina, 177, 180, 186, 190, 191,
    199, 203, 207
Argentine Meteorological Service, 191
Aristotle, 228
Aristotelian "meteoric tradition," 35
Armstrong, John, 12
Arnold, David, 2
Arnold, Hap, 144
Arrhenius, Svante, 233, 235
Asia, 58, 229

Assmus, Alexi, xv, 94, 95, 103, 106, 108
Atlantic, the, 70, 85, 150, 179, 184, 194
Atlantic Basin, 188
Atlantic storms, 6
Atlas des Orages, 81, 82
Auslandsdeutschen investors, 181
Australia, 75, 176
Austrian Empire, 124, 129, 181
Austria, First Republic (1918–1934), xvii, 117, 188, 120–130, 132, 135
Austrian Alps, 127, 128, 134, 135
"Austrian climate," 119
Austrian Meteorological Society, 116
Automobile Club of Southern California, 151

Babbage, Charles, 42 43, 44
Bacon, Francis, 47
Baden, 122
Baily, Francis, 51
Barker-Benfield, G. J., 9
Barkey, Karen, 134
Barrett. W. F., 234
Basic Principles of Weather Forecasting (Starr), 159
Batavia, 204, 207
Bates, Charles, 150
Bath, 14
Bauer, Peter Paul von, 181, 184, 189, 200
Bauhaus movement, 181
Beagle voyage, 74
Beaufort, Francis, 57
Beaufort scale of wind strengths, 125, 162
Beck, Lewis, 231
Beddoes, Dr. Thomas, 16
Bendix Aviation, 191
Ben Nevis, 107
Ben Nevis Observatory in Scotland, 94
Bergen, Norway, xvi, 148, 194
Bergen Geofysisk Institut, 192

Bergen school, xvii, 142, 148, 149, 151, 153, 155, 158, 160, 177, 193–195, 197, 198, 201, 204, 205–207
Bergeron, Tor, 149
Berghaus, Heinrich, 70
Berlin, 191
Bessel, Friedrich Wilhelm, 51
Bioklimatische Beiblätter, 122
Birkbeck, George, 5
Birt, William Radcliffe, 36, 50, 52, 53
Bjerkes, Hedvig, 193
Bjerknes, Jacob, 144, 145, 148, 164, 192–195, 197, 198, 201, 202, 207
Bjerknes, Vilhelm, xvi, xvii, 132, 144, 148, 149, 191, 192, 205
Bjerknes method, 148
Black, Joseph, 39
Blair, Tony, 227
Bloxam, James Mackenzie, 22
Blue Hill Observatory, 179
Bodin, Jean, 228
Bolívar, Simon, 175
Bolivia, 181, 203
Borda's repeating circle, 43
Boston, 152, 179
Boulanger, 6
Bowman, Isaiah, 154
Boyle, Robert, 37
Boys, C. V., 99
Bradford Grammar School, 102
Brandes, Heinrich, 69, 70, 71
Brazil, 177, 178, 181, 188, 189, 190, 191, 199, 200, 203
Breithorn, 43
Brezina, Ernst, 124, 126
Britain, 3, 18, 57, 58, 70, 72, 75, 78, 86, 117
British Association Committee on electrolysis/electrical standards, 100
British Association for the Advancement of Science, 50, 57, 71
British Commonwealth, 197
British Empire, 179

British Imperial Airways, 194
British Isles, 79
British Meteorological Office, 77, 99
Brunngraber, Rudolf, 135
Brussels, 72, 75
Buddha, x
Buenos Aires, 179, 190, 198
Bulletin Meteorologique International, 81
Bulletin of International Meteorology, 83, 85
Bureau of Aeronautics (U.S. Navy), 154
Burgess, Thomas, 22
Buys Ballot, C.H.D., 82
Byers, Horace, 151, 152, 155, 156, 163, 165

Cabinet Cyclopaedia, 48
Caius College, 102
Calcutta, 51, 57
California, 151, 192
California Institute of Technology (Cal Tech), 153, 157, 190
California State forester, 151
Callendar, G. S., 227, 236, 237
"Callendar Effect," 237
Cambridge, England, 240
Cambridge Analytical Society, 54
"Cambridge Attitude Towards Meteorology, The" (Shaw), 96
Cambridge-MIT Institute, 240
Canada, xvii, 164, 175–177, 193, 194
Canada, Government of, 238
Canadian Meteorological Service, 179–180, 194
Cannon, Susan Faye, 51
Cape Horn, 175
Cape Observatory, 51
Cape of Good Hope, 51, 52, 55
Cape Town, 55, 57
Caribbean, the, 177, 184, 186, 205
Carinthia, 117
Catania, Sicily, 45
Cavendish Laboratory, 93–96, 98, 100, 103, 107, 108

Central America, 186
Central Europe, xvii, 115, 120, 128, 132, 135, 136
Central Institute for Meteorology and Geophysics (Vienna), 116, 117, 119, 121, 123, 134, 144
"Chaco War," 199
"Changing Atmosphere: Implications for Global Security, The," 238
Charney, Jule, 165
Cheyne, Dr. George, 9, 11
Chicago, 179, 207
Chile, 178, 190
Chimborazo, 175
Chisholm, Colin, 20
Civil Aeronautics Act of 1938, 156
Clark, James, 20
CLEAR, 117
Clemenceau, George, 117
"Climate of the Alps, The," 128
"Climatology of Austria, The," 119
"Climatology of Bukowina, The," 119
Clutterbuck, Henry, 5
Cockburn, William, 1
Coen, Deborah, xvi, 115, 144, 181
Cold War, 203
Coleman, William, 21
Colombia, 184
Colombian Amazon, 181, 184
Committee on Physics and Meteorology, 46, 52
Compton, Arthur, 157
Compton, Karl, 154
Concepción, Chile, 178
Condor Syndikat (Syndicato Condor), 184, 190, 191
Conrad, Victor, 119, 120, 121
Coolidge, Calvin, 152, 193
Copenhagen, 81
Coulomb, Charles Augustin de 47
Cowper, Lady, 14
Cowper, William, 14
Crimea, 81
Crutzen, Paul, 227, 242
Cuba, 175, 181, 184, 186
Cuban National Observatory, 184

Cuban Revolution (1959), 203
Cuevas, Rafael Dávila, 202
Cullen, William, 9
Cushman, Gregory, xvii, 175, 144, 164
Czechoslovakia, 122

Daily News (London), 71
Dalton, John, 39
Daltonian atomic events and force laws, 41
Daniell, John Frederick, 5, 35
Darwin, Charles, 74
Darwin, George, 99
Davos, 120
DC-series of aircraft, 190
Department of Agriculture, 154, 156
Depot of Charts and Instruments, 74
Derbyshire, 6
Dessau, 181
Deutsche Luft Hansa, 181, 199
Deutschen Petroleum AG 181
Deutsche Seewarte (German Marine Observatory), 184, 186, 187, 188, 191, 200
Diderot, Denis, 6
Dorno, Carl, 120
Douglas Aircraft, 190
Dove, Heinrich Wilhelm, 35, 50, 69, 70
Drama of Weather, The (Shaw), 94
Du Bos, Jean-Baptiste, 228, 229
Dunbar, James, 11
Dutch East Indies, 204

Earth, 37, 38, 39, 40, 48, 50, 53, 58, 141, 223–226, 232, 233, 235, 236, 238, 240
East Africa, 179
Eastern Europe, 118, 133, 177
Ecuador, 177, 199, 200, 201, 203–205
Edinburgh University, 1, 8, 9
Egypt, 233
Eighth American Scientific Congress, 198

Ekholm, Nils, 235
El Niño, 189, 201, 204
Elster, 106, 107, 108
Emmanuel College, 96
Encyclopaedia Britannica, 40, 52
England, 71, 72, 118
English Malady, 18
Enlightenment, the, xiv, 7, 12, 21, 227
Espy, James, 49, 72, 75
Essay on Regimen for the Preservation of Health (Adair), 14
Europe, xi, xiii, 2, 12, 43, 46, 53, 54, 58, 69, 75, 81, 82, 118, 122, 129, 175, 178, 191, 195, 198, 203, 228–230, 232, 240
European continent, 70
European Hippocratism, 20, 21
Exner, Felix, 117, 119, 120, 128, 132
Experimental Weather Reporting Service, 151

Falconer, William, 6, 12
Feldman, Theodore, 35
Ferrel, William, 232
Ficker, Heinrich, 120, 128, 129
First World, 208
First World War, xi, xvi
Fitzgerald, Emily Mary (Lady Kildare), 16
FitzRoy, Robert, 72–77
Fleming, James Rodger, xviii, 35, 223
Fletcher, Joseph, 243
Florida, 186, 200
Florida State University, 142
Focke-Wulf, 190
Föhn winds, xiii, 130
Forbes, James David, 45, 46, 51, 54, 57
Forry, Dr. Samuel, 232
Forster, Thomas Ignatius Maria, 5
Foucaldian heterotopias, 13
Foucault, Michel, 2
Fourier, Joseph, 44, 46, 233, 235
Frankland, Dr. Edward, 234
Franklin, Ben, 7

France, 9, 46, 69, 78, 81, 86, 117, 118, 176, 178, 180
French Academy, 228, 229
French Academy of Sciences, 233
Fresnel, Augustin, 46
Friedman, Robert Marc, xvi, 144, 149
Fron, E., 81

Geceta de Méjico, 178
Galápagos Islands, 199, 200, 201
Galen, 9
Galison, Peter, xv, 94, 95, 99, 103, 106, 108
Galmarini, Alfredo, 199
Galton, Francis, 76, 78, 79, 80, 82
Galveston, Texas, 175
Gauss, Karl Friedrich, 56
Gay-Lussac, J. L., 39
Geiger, Rudolf, 133
Geitel, H., 106, 107, 108
Gemmellaro, Mario, 45
General Land Office, 232
Geological Society, 38
Geopsyche (Hellpach), 133
Georgian England, 13
Georgics (Virgil), 11
Georgii, Walter, 186
German Alps, 120
German Weather Service, 192
Germany, 46, 82, 117, 126, 133, 177, 180, 181, 184, 187, 191, 195, 199, 200
Glaisher, James, 71, 72
Glazebrook, Richard Tetley, 97, 98, 99, 100, 102
Global Climate and Ecology Institute, 242
Gold, E. E., 99
Good, Gregory, xiv, xv
Good Neighbor Policy, 198
Gorczynski, Wladyslaw, 198, 201, 204
Gordon, Thomas, 10
Gore, Al, 243
Grace Lines, 189
Graf Zeppelin, 186, 188

Graz, 129
Great Britain, 180, 195
Great Depression, 178, 189, 190, 194
Great Exhibition (1851), 80
Great Exhibition in London (1851), 72, 80
Great Powers, 176
Great War, xvii, 117, 119, 124, 135, 154, 184
Greenland, 186
Greenwich, 85
    Observatory, 71, 72, 107
Gregg, Willis, 155
Gregory, James, 9, 12
Gregory, John, 9
Guadalcanal, 205
Guantánamo Bay, 205
Guatemala, 186, 201
Guayaquil airport, 202
Guggenheim, Harry, 145, 150, 151, 152
Guggenheim Fund, 146, 150, 152, 153
    promotion of Aeronautics, for the, 192
Guildford, 13
Gulf of Genoa, 101
Gulf of Mexico, 184

Habermasian public spaces, 13
Hadley, George, 48
Haiti, 176
Hales, Rev. Stephen, 8
Halley, Edmond, 37, 48, 69
Hamburg, 184
Hann, Julius, 119
Hanoi, 204
Hansen, James, 223, 224, 243
Habsburg Empire, 116, 122
Hare, Robert, 49
Hargrave, Lawrence, 179
Harris, Steve, x
Hatfield, Charles M., 147
Haushofer, Albrecht, 133
Hauslab, Josef von, 79
Hausmann, Walter, 120–121
Havana, 178, 184, 188, 207

Heimatschutz conservation movement, 127
Helland-Hansen, Bjørn, 197
Hellman, Gustav, 69
Hellpach, Willy, 133
Helmholtz, Hermann Ivon, 132, 153
Henderson, Thomas, 51, 54
Hennen, John, 5
Henry, Joseph, 35, 72, 75, 85
Herschel, John, xiv, xv, 35–58, 71, 74
Herschel, William, 50
Hess, Seymour, 141, 142, 164
Hessenbruch, Arne, 106
Hildebrandsson, Hugo, 70, 82
Hill, Britton Armstrong, xiii
Hindenburg disaster, 189
Hippocratic corpus, 228
Hippocratic medicine, 7
Hippocratic "places," 8
Hippocratic prescription, 244
Hitler, Adolf, 136
Hobart, 57
Hoffman, Ross, 241
Hoffmeyer, Niels, 81
Hofmannsthal, Hugo von, 127, 135
Högbom, Arvid, 235
Holmboe, Jörgen, 192, 197 204
Hooke, Robert, 37
Hoover, Herbert, 152, 193
Hope, Thomas, 14
Howard, Luke, 5, 43
Huddersfield, 20
Hudson, James, 48–49, 56
Humboldt, Alexander von 39, 44, 56, 69–71, 175
"Humboldtean medicine," 121
Hume, David, 9, 11, 229, 230
Hunter, John, 9
Huntington, Ellsworth, xii
Hutchison, Senator Kay Bailey, 241
Hydrographic Office of the Admiralty, 74

IMO Aerological Commisssion, 195
Imperial College of Science and Technology, 195

India, 2, 179, 203
India Civil Service, 96
Institute of Meteorology (University of Chicago), 156
Institute of Tropical Meteorology (University of Puerto Rico), 160, 204
Instructions for Making and Registering Meteorological Observations (Herschel), 56
Inter-American Institute of Meteorology, 202
Interdepartmental Committee for Cultural and Scientific Cooperation, 202
International Geophysical Year (1957–1958), 204
International Meteorological Organization (IMO), x, 191
IPCC Second Assessment Report (1995), 238
IPCC Third Assessment Report (2001), 227
Isère, 233
Israel, Yuri, 242
Italy, 46, 180
Ito, Kenji, 145

Jamaica, 184
Jankovic, Vladimir, xiv, 1, 35, 36, 50, 52, 125
Japan, 145, 203
Java, 179
Jefferson, Thomas, 230, 231
João I, 178
Johnson, James, 20
Johnson, Samuel, 11, 12
Jordanova, Ludmilla, 2
Journal of Meteorology, 141, 163, 164
Junior Observer, 148
Junkers Flugzeugwerke AG, 181, 189, 190
Jupiter, 48
Jurin, James, 5

Kaiser, David, 145
Kames, Lord, 9, 11

Kämtz, Ludwig, 46
Kew Observatory, 85, 107
Key West, 188
Kingston, 184
Kohler, Robert E., 99
Kopenhagener Geist, 145
Köppen, Wladimir, 184
Kosmos, 39
Kraus, Gregor, 129

La Pérouse, 178
Laplace's theory of ocean tides, 153
Lapland, 179
Latin America, xvii, 160, 164,
    175–180, 186, 189, 193, 198,
    200, 201, 203
"Law of Rotation" (Dove), 50
Leeward Islands, 13
*Les mouvements generaux de
    l'atmosphere*, 81
Le Verrier, Urbain, 53, 80, 81, 83
Lima, 202
Lindbergh, Charles, 145, 150, 180,
    189
Linke, Friedrich, 122
Lisbon, 178
Lit and Phil philanthropists, 19
Lloyd, John Augustus, 56, 57
Lloyd Aéreo Boliviano (LAB), 181
Lockheed Electras, 194
Loening hydroplanes, 188
London, 14, 71, 77, 85, 234
Loomis, Elias, 69
Loos, Adolf, 127
Lorenz, Edward, 165
Los Angeles, 151, 197, 200, 207
Lovelock, James, 243
Luftwaffe, 188
Lyell, Charles, 38

Mackenzie, James, 12
Maclear, Thomas, 55
"Macro-engineering options for
    Climate Change Management and
    Mitigation," 240
Madeira, 22

Madero, Francisco, 179
Magic Mountain (Mann), 120
Magnetic Crusade, 55, 57, 193
Manchester, 19
Manila, 204, 207
Mann, Thomas, 120
Manual of Meteorology (Shaw), 94,
    108
Manual of Scientific Enquiry
    (Herschel, editor), 55
Margules, Max, 132
Marie-Davy, E. H., 81
Mariotte, Edmé, 37, 39
Marquard, Odo, 11
Mars, 141, 164
Marshall Plan, 203
Marsigli, 37
Marvin, Charles F., 147, 148, 149,
    150, 153, 154
Mason, John Abraham, 22
Massachusetts Institute of Tech-
    nology (MIT), 145, 152, 153,
    154, 157, 192
Mauritius, 56
Maury, Matthew Fontaine, 74, 75,
    76, 77
Maxwell, James Clerk, 96, 96, 100
Mayan ruins, 189
Medellín, Colombia, 202, 203, 207
Medical Cautions (Adair), 14
Mediterranean, the, 2, 5, 6, 228
Melloni, Macedonio, 46
Mena, Luis, 203
Mesoamerica, 189
Meteor expedition (1925–1927),
    186
Meteorographica (Galton), 78, 79
Meteorological Committee of the
    South African Literary and
    Philosophical Society, 55
Meteorological Council of the Royal
    Society, 93, 96, 100
Meteorological Service of Ecuador, 201
Meteorological Society of London, 5
Meteorological Society of the
    Palatinate, 69

Meteorologische Zeitschrift, 116, 120, 122
"Meteorology" (Herschel), 40, 50
Mexicana, 189
Mexican Meteorological Service 198, 201
Mexico, 178, 181, 184, 190, 203
Mexico City, 150, 201
Miami, 207
Milan, 41
Millás, José Carlos, 184, 191
Millikan, Robert, 154
Ministry of Education, 123, 131
Ministry of Marine, 81
Ministry of Sanitation and Health, 124
Mitchell, Gen. Billy, 144, 145, 150, 154
Mitchell, Charles, 154
Mitman, Gregg, 121
Moffett, Adm. William, 154
Mönichkirchen, 121
Monroe Doctrine, 186
Monte Rosa, 43
Montesquieu, Baron de (Charles de Secondat), 6, 228, 229
Montevideo, 179, 180, 191, 198
Montevideo meteorological conference, 191
Montgolfier brothers, 178
Montpellier, University of, 8
Montreal Protocol, xii
Montsouris, 81
Morandi, Luis, 179, 180, 199
Mount Etna, 44, 45
Murchison, Roderick, 58
Musil, Robert, 134, 135
Myer, General Albert, 85

Namias, Jerome, 152, 155
Napoleon, 233
Naswetter, Dr. Erwin ("Dr. Wetweather"), 121
NASA, 142, 223
NASA's Institute for Advanced Concepts, 241

National Academy of Sciences, 165, 238, 239
National Physical Laboratory, 99
National Socialism
    National Socialists, 133, 199
    National Socialist trade policy, 190
Nazi imperialism, 133
Natural History of Body and Mind (Adair), 13–14
Nature, 77, 107
Netherlands, The, 180
Neumayer, G., 82
Newall, H. F., 98, 102
Newfoundland, 194
Newton, Isaac, 37
New World, xvii, 227, 229, 230
New York City, 188
New York, state of, 232
New York University, 153, 157
New Zealand, 203
New Zealand Meteorological Office, 204
Nobel Prize, 94
Norfolk fen, 6
North Africa, 200
North America, 54, 70, 177, 192, 195, 197, 228, 230
North Atlantic, 186
Northern Hemisphere, 237
North Pole, 85
Northumberland, Duke and Duchess of, 14
Norway, 160, 193, 197
"Norwegian School," 192, 195, 198
Notes on Virginia (Jefferson), 230
Notgemeinschaft der Deutschen Wissenschaften, 125

Oakland, 151
Oceania, 179
"Of the Populousness of Ancient Nations" (Hume), 229
Ostfriesland (German battleship), 150
Ottoman empire, 122
Outlines of Astronomy (Herschel), 49, 50

Pacific Coast, 189, 197, 200
Pacific Ocean, 54, 178
Palmer, Clarence, 204, 205, 207
Panair do Brasil, 199
Panama Canal, 186, 193, 200
Panama Canal Zone, 184, 208
Pan American Airlines, 160, 188, 189, 194, 199, 200
Pan American-Grace Airways (Panagra), 200, 202
Pan-American regional identity, xvii
Paraguay, 199
Paris, 41, 44, 85, 188
Paris Observatory, 69, 70, 76, 81, 83
Pascal, Blaise, 37
Pascala, Stephen, 243
"Pass of the Jura at Champagnole Montagne Cornice" (Herschel), 43
Pearl Harbor, 157, 202
Peru, 199, 202, 204, 205
Petterssen, Sverre, 192, 194
Philosophical Transactions of the Royal Society, 106
"Physical Geography" (Herschel), 40
Physical Geography of the Sea (Maury), 75
Piddington, Henry, 50
Pitcaine, Dr. Archibald, 1
Poisson, Siméon Denis, 47
Poland, 195
Policy Implications of Greenhouse Warming, 239
Politics (Aristotle), 228
Porter, Roy, 11
Portugal, 176
Pouillet, C.-S.-M. 46
Practical Physics (Glazebrook and Shaw), 97, 98
Prandtl, Ludwig, 132
Preliminary Discourse on the Study of Natural Philosophy (Herschel), 37–39, 43, 47, 55
President's Science Advisory Committee, 238
Prevost, Pierre, 39
Price, Richard, 19

Priestley, Joseph, 15
Principles of Geology (Lyell), 38
Prinsep, James, 51, 57
Public Health Office of the Federal Ministry of Social Administration, 121
Puerto Rico, 178, 204, 207, 208
Putin, President, 242
Puy-de-Dôme, 45

Quito, 204
Quito Astronomical Observatory, 203

Rabinbach, Anson, 127
Ratzel, Friedrich, 133
Rayleigh, Lord, 97, 100
Redfield, William, 49, 50
Red Vienna, 118, 124
Reed, Charles D., 154
Réflexions critiques sur la poësie et sur la peinture (Du Bos), 228
Reicheldefer, Francis, 142, 144, 145, 150, 155, 156, 164, 193
Reichswetterdienst, 188
Reid, William, 50
"Remarques générales sur les temperatures du globe terrestre et des espaces planétaire" (Fourier), 44
Restoring the Quality of Our Environment, 238
Revenge of Gaia, The (Lovelock), 243
Rhône, 233
Richardson, Lewis, 132, 159
Riehl, Herbert, 205, 207
Riley, James, 2
Rinke, Stefan H., 181
Rio conference in 1992, 227
Rio de Janeiro, 179, 189, 190 191, 200
Rio Grande, 188, 207
Ritchie, William, 45
Rockefeller, Nelson, 199, 201, 202
Rocky Mountains, 128
Romanov empire, 122
Roosevelt, Franklin D., 153, 154, 155, 157, 198

Ross, James Clark, 57
Rossby, Carl-Gustaf, xvii, 142, 144–146, 148–153, 155, 156, 160, 162, 165, 192, 193, 195, 197, 201, 202, 204, 207
Royal and American Meteorological Societies, 194
Royal Astronomical Society, 54
Royal Charter storm of 1859, 72, 73, 74
Royal College of Physicians of Edinburgh, 13
Royal Institution, 233, 234
Royal Medical Society, 13
Royal Meteorological Society, 92, 101
Royal New Zealand Air Force, 205
Royal Observatory, 55
Royal Society, 46, 54, 56, 107
Royal Society barometer, 51
Rumford, Count, 19
Rupke, Nicolaas, 121
Ruskin, John, xi
Russia, 46, 129
Rutherford, Ernest, 106
"Rutherford and the Meteorologists" (Hessenbruch), 106
Ryan, Edward, 51, 57

Sabine, Edward, 39, 71
Sabine, Elizabeth Leeves, 39
Salzburg, 123, 127
Salzkammergut, 123
Samoa, 179
San Francisco, 223
San Francisco Bay area, 151
San Juan, 205
Santa Elena peninsula, 200
Santa Monica, California, 190
Santiago de Chile, 203
Saturn, 48
Scheele, C. W., 39
Schmidt, Wilhelm, 119, 122, 123, 124, 125, 126, 129, 131, 132
Schuster, Arthur, 99

Science Advisory Board, 154, 155
"Scientific Basis of Modern Meteorology, The" (Rossby), 156
Scirocco, 3
Scotland, 71
Scottish Enlightenment, 9
Scripps Institution of Oceanography (SIO), 197, 198
Searle, 102
Second Latin American Scientific Congress, 179
Secretary of Agriculture, 147
Secundus, Monro, 9
Seilkopf, Heinrich, 186, 187, 188
Shammas, Carole, 14
Shaw, Napier, 93, 95–104, 106, 108, 109
Short, Thomas, 5
Siberia, 179
Sigmond, George, 1
Sixth Pan American Conference, 188
Smith, Adam, 9
Smithsonian Institution, 72, 75, 76, 85, 147, 193
Sociedad Colombo-Alemana de Transportes Aéreos (SCADTA), 181, 184, 186, 189, 199
Socolow, Robert, 243
Sonnblick observatory, 128
South Africa, 51, 55, 176
South America, 54, 75, 177–179, 186, 191, 199–201, 207, 208
South Atlantic, 186
Southern cone, 181, 201
Southern Hemisphere, 203
Southern Oscillation, 204
South Pacific, 160, 203
South Pole, 186
South Tyrol, 117–118, 120
Spain, 176
Spanish America's Enlightenment, 178
Spilhaus, Athelstan, 152, 153
Spirit of St. Louis, The, 188
Spitsbergen, 179

Staley, Richard, xv, xvi, 93
Starr, Victor, 159, 160
Steinhauser, Friedrich, 124
Stewart, George R., 155
Stockdale, Percival, 12
Storm (Stewart), 155
Sudetenland, 117
Sun, 39, 53, 54
Sverdrup, Harald, 197, 198, 204
Swan Island, 205
Sweden, 144, 149, 192
Swedish-American Foundation, 192
Switzerland, 180
Sydney Observatory, 179
Symons, George James, 70, 85
Syndicato Condor, 200

Tatem, James, 5
Taylor, Alfred Swaine, 50
Taylor, G. I., 132
Teisserenc de Bort, Léon, 70, 82
Teller, Edward, 241
Texas, 241
Theodore, Adolfe, 178
Theophrastus, 228
Third Reich, 133, 135
Third World, 203, 208
Thomson, Andrew, 194
Thomson, J. J., 95, 97, 98, 102, 103, 105, 108
Time magazine, 224
Toronto, 57, 193, 194
Torrid Zone, 204
Transactions of American Philosophical Society, 69
Trans-Canada Air Lines (TCA), 194
Transcontinental and Western Airlines (TWA), 190
"Transportation Is Civilization," 179
Traweek, Sharon, 145
Treatise on Astronomy, A. (Herschel), 48, 49
Treaty of St. Germain (1919), 117
Turner, H. H., 99
Turner, Roger, xvii, 141, 193, 197

Tyndall, John, xii, 233–235
Tyndall Centre for Climate Change Research, 240, 241
Tyrol, the, 123

Ukraine, 116, 119
United Airlines, 194
United Kingdom, 208
United Nations, 238
  Framework Convention on Climate Change, 242
United States, xiii, xvii, 70, 72, 75, 78, 86, 129, 144, 164, 176, 178, 186, 188, 190, 191, 193, 195, 197, 200–202, 204, 207, 208, 224, 240
  Army, 199
  Army Air Force, 141, 145, 151, 154, 161, 188, 200, 203
  Army Medical Department, 232
  Army Signal Service/Corps, 70, 83, 85, 193
  Department of Commerce, 186
  Navy, 141, 151, 152, 154, 191
  Navy Aerology section, 144, 150
  Pentagon, 240
  Postal Service, 188
  Sixth Weather Squadron, 200
  War Department, 186, 188
  Weather Bureau, 142, 144–156, 184, 191–193, 195, 197, 198, 201–203
Universidad Nacional Mayor de San Marcos, 202
University Meteorological Committee (UMC), 156, 157, 160, 161, 162, 164
University of California, Los Angeles (UCLA), 153, 157, 197, 201, 202, 207
University of Cambridge, 93
University of Chicago, 142, 153, 156, 157, 204, 207
University of Kyoto, 202
University of Stockholm, 149

University of Toronto, 194
University of Vienna, 128
Uruguay, 180
USS Utah, 156

Vanity Fair, 224
Vargas, Getúlio, 189, 200
Venezuela, 181, 184, 190
Versailles Treaty, 180
Vienna, xvi, 118, 124, 126, 128, 130, 144
Virgil, 11
Vonnegut, Bernard, xii

Walker, John, 20
Warner, Edward, 152
Washington, D.C., 72, 150, 192, 193, 198, 201
Weather Book (FitzRoy), 73
Weather Forecasting in the United States, 147
Weather Modification Research and Technology Transfer Authorization Act of 2005, 241
Wegener, Alfred, 184
Weightman, R. Hanson, 201
Weimar government, 122, 181
West Africa, 186
Western Air Express, 151

Western Hemisphere, 175, 181, 186, 187, 192, 198
Wexler, Harry, 152, 155
Whewell, William, 50, 51, 55
Whitnah, Donald, 147
Whytt, Robert, 9
Willet, Hurd C., 152
Williamson, Hugh, 231
Willich, Anthony Florian Madinger, 17
Wilson, C. T. R., 93–95, 98–100, 102–109
Winnipeg, 193
World Meteorological Organization, 164
World Meteorological Programme, 238
Worlds in the Making (Arrhenius), 235
World's Columbian Exposition, 179
World War I, 128, 132, 144, 148, 150, 177, 180, 191, 193
World War II, xvii, 141, 145, 164, 177, 190, 199, 204
Wray, Gustavo, 202
W. R. Grace & Co., 189
Wright Brothers, 178, 179

Yearbook of Agriculture: Climate and Man (1941), 156

Zimmerman, Col. Donald, 157